T0260659

OBSERVING DARK INNOVATION

After Neoliberal Tools and Techniques

Ryan T. MacNeil

BRISTOL
UNIVERSITY
PRESS

First published in Great Britain in 2024 by

Bristol University Press
University of Bristol
1–9 Old Park Hill
Bristol
BS2 8BB
UK
t: +44 (0)117 374 6645
e: bup-info@bristol.ac.uk

Details of international sales and distribution partners are available at bristoluniversitypress.co.uk

British Library Cataloguing in Publication Data
A catalogue record for this book is available from the British Library

ISBN 978-1-5292-3119-9 hardcover
ISBN 978-1-5292-3120-5 ePub
ISBN 978-1-5292-3121-2 ePdf

Cover design: Lyn Davies
Front cover image: mauritius images GmbH / Alamy Stock Photo
Bristol University Press uses environmentally responsible print partners.
Printed and bound in Great Britain by CPI Group (UK) Ltd, Croydon, CR0 4YY

FSC
www.fsc.org
MIX
Paper | Supporting
responsible forestry
FSC® C013604

Contents

List of Figures and Tables

Figures

Tables

Acknowledgements

First and foremost, this book was possible thanks to the love and encouragement of my wife Amanda and my children Brennan and Emily.

The beginnings of this work go back ten years and, unfortunately, I do not have the space to thank everyone who supported me since then. However, specific thanks are owed to those who read drafts of this writing: Kendra Carmichael, Larry Corrigan, Nick Deal, Susie Decoste, Gabie Durepos, Jean Helms Mills, Kevin Laing, Albert Mills, Donna Sears, Terry Weatherbee and Kristin Williams. Faculty and students in Acadia University's social and political thought programme provided valuable feedback on Chapters 3 and 7, especially Andrew Harding and Ian Wilks. D.J. Decoste helped me think about quantum physics. Brian Sanderson and Anna Redden helped me understand technical details from their published work, which I discuss in Chapter 6. I am grateful for research support provided by Rosemary Barbour at the Nova Scotia Archives, Ron Houlihan at the Patrick Power Library, Saint Mary's University, and my colleague Britanie Wentzell at the Vaughan Memorial Library, Acadia University. Eric von Hippel astounded me with his quick and gracious reply when I wrote to ask about one word in his 1988 book. I was starstruck during the virtual coffee conversation that followed and floored again when he agreed to let us publish part of his email response in Chapter 2. Paul Stephens at Bristol University Press was tremendous in helping to focus and shepherd this book from idea to publication, and I am grateful to the whole Bristol University Press team, including the seven anonymous reviewers, for their efforts. Finally, I would not have persisted in writing this book without the coaching provided to me by John Guiney Yallop.

The empirical material discussed here comes from my PhD research, which was supervised by Claudia De Fuentes, examined by Roman Martin, and benefitted immensely from committee member Ryan Gibson. My approach changed considerably from there to here, partly thanks to paper rejections from two decidedly mainstream innovation journals, and partly thanks to constructive conference feedback at the European Group of Organizational Studies (Athens, 2015), the International Schumpeter Society (Montreal, 2016), International Critical Management Studies (Milton Keynes, 2019), the Society for the Social Studies of Science (Cholula, 2022 and Honolulu, 2023) and the Atlantic Schools of Business (Halifax, 2014 and Wolfville, 2022). This research was possible thanks to the financial support of a Joseph Armand Bombardier Doctoral Scholarship from the Social Sciences and Humanities Research Council of Canada and the Rath Professorship in Entrepreneurship at Acadia University.

1

Dark Innovation

Since the 1930s, physicists have known that a great deal of matter is missing from their observations (de Swart et al, 2017). Their calculations tell us that all the matter we can currently observe *must* be only a fraction of the matter in the universe. But even the best scientific instruments of the day cannot directly observe dark matter.

We have a similar problem with research on innovation. In laying out the greatest challenges for the future of innovation studies, Ben Martin (2013, 2016) argued that a great deal of potentially interesting phenomena remain in the dark. But unlike physics, where an average of three new papers per day are focused on the elusiveness of dark matter (de Swart et al, 2017), hardly anyone is systematically working to reveal dark innovation. Yes, we know a good deal about several types of innovation that tend to be overlooked, downplayed, and marginalized: services innovation (for example, Gallouj and Weinstein, 1997), public innovation (for example, Røste, 2005; Windrum and Koch, 2008; Mazzucato, 2013), and user innovation (for example, von Hippel, 1986). However, research on innovation still tends to fixate on particular classes of technology (computers, biotechnology, etc.) (Martin, 2013, 2016). Other innovation *matters* – some might be harmful and some might be critical to our survival.

This book asks why we struggle to observe dark innovation. I argue that our research tools and techniques – our 'scientific instrumentalities' (de Solla Price, 1984) – were built with only certain forms of innovation in mind. They conceal as much as they reveal. As John Law might tell us, any 'method assemblage' (2004, p 14) will enact both presence and absence. In short, science is political (Polanyi, 1962) and those politics are scripted (Winner, 1980; Latour, 1992; Akrich, 1994) into our scientific instruments. The trouble is that scholarly norms tend to privilege the epistemic outcomes of science – the knowledge or ideas we produce. We tend to separate these from the *techne* – the tools and techniques – that allow for knowledge production. History brushes over the new instruments that enable scientific breakthroughs (Hughes, 1976).

It is therefore not surprising that we have forgotten how the study of scientific instruments enabled multiple breakthroughs in innovation theory. Key figures in the history of innovation studies – Eric von Hippel (1976, 1986), Christopher Freeman (1974), Stephen Kline (1985; Kline and Rosenberg, 1986), and Nathan Rosenberg (Rosenberg, 1982; Kline and Rosenberg, 1986) – all featured scientific instruments in their theory development. Their contributions are among the top ten most-cited works in innovation studies (Fagerberg et al, 2012a). But innovation in scientific instruments is hardly mentioned in present-day research. This book is a reminder of the important societal role of scientific instrument innovation, but it is not an attempt to create another top-ten theoretical contribution. Rather, it is a call for critical scholars to help retool the instrumentalities of innovation research. It is my contribution to the agenda set by Benoît Godin and Dominique Vinck 'to develop a research program on *Critical Studies of Innovation*' (emphasis in original) (2017a, p 12).

The challenge

Let me begin by explaining how 'dark innovation' fits into a broader critical studies of innovation agenda. To date, no one has directly confronted Ben Martin's (2013, 2016) version of the dark innovation challenge. Martin's challenge is mentioned briefly by Alf Rehn and Anders Örtenblad (2023) in the introduction to their edited collection *Debating Innovation*. But the only direct inquiry was a special issue of *Industry and Innovation* that used 'dark innovation' as a euphemism for the negative dimensions of innovation (Coad et al, 2021). This was an important acknowledgement that innovation sometimes has nefarious intentions and/or outcomes. Personally, I have been disturbed by history books on companies like IBM (Black, 2001), DuPont (Ndiaye, 2007), and I. G. Farben (Hughes, 1969). These books describe how such companies gained strength by developing tools for mass murder. And so, I am worried about how 'bad' innovation is so often ignored by innovation research. There is an undoubtable pro-innovation bias (Godin and Vinck, 2017a) where scholars, policy makers, and everyday citizens ignore, marginalize, and brush past the innovation that is bad for us and bad for this planet. But I come at this issue sideways. In this book, I try to capture 'bad' innovation – and more – under the umbrella of 'dark innovation'. I take this broader view because innovation studies also neglects many innovation activities that are good for us. I am interested in the many varieties of innovation – and 'novation' (Godin and Vinck, 2017a, p 3) – that are cast into the dark. And so, my approach expands on Ben Martin's (2013, 2016) dark matter metaphor. Dark innovation could be anything absent from our observations.

In this book, I use the example of public innovation in physical goods. Prior research on public innovation (see a review in de Vries et al, 2016) has ignored, dismissed, or minimized innovation in goods. With few exceptions (for example, Bugge et al, 2011; Arundel and Huber, 2013), there is a common rhetorical pattern: first, deploy anecdotes about novel public sector inventions (such as the Global Positioning System [GPS]) to convince readers that public organizations innovate; next, ignore public innovation in goods and either characterize the state as a facilitator of market-based innovation (as in Mazzucato, 2013, 2016) or shift focus to any public innovation aside from physical goods (as in Walker et al, 2002; Halvorsen et al, 2005; and Koch and Hauknes, 2005). For example, a prominent book in the field begins with a chapter that mentions public innovation in medical technologies, medical instruments, and pharmaceutical products (Windrum and Koch, 2008). The same book concludes with the surprisingly definitive counterclaim that 'technological innovations, especially goods, are the exclusive domain of the private sector' (Windrum and Koch, 2008, p 239). It has been said that the 'appropriate' way to understand product innovation in the public sector is by focusing on services (Walker, 2014, p 23). Public innovation in goods is considered inappropriate. Therefore, it is sidelined from innovation studies.

This prejudice is both political and methodological. Throughout this book, I show how common methodological tools and techniques carry the 'neoliberal bias' (Fløysand and Jakobsen, 2011; Cooke, 2016), neoliberal 'dogma' (Lundvall, 2016), 'market bias' (Gallouj and Zanfei, 2013; Cruz et al, 2015) or 'market ontology' (Pfotenhauer and Juhl, 2017) that dominates innovation studies. This is not quite the full-blown 'phobic reaction to the state' shared by early proponents of neoliberalism (Peck, 2010a, p xii). However, it is clearly a variation on the neoliberal theme: there is a sense that government should be constrained in favour of private companies. What is interesting is how this theme plays out in innovation research. I do not approach neoliberalism as a coherent concept or use it as 'a heavy-handed tool of social analysis that knows its answer in advance' (Phelan, 2014, p 2). But I do think of it as a grand narrative – one that is 'somewhat ambiguous and situationally specific' (Peck et al, 2018, p 4). My interest is in specific ways in which mainstream innovation studies constrain observations of public sector innovation.

Along the way, I am making a broader point. I am exploring the neoliberal bias against public innovation because it is one example of the 'dark matter' that lies beyond the edges of innovation studies. This is where I depart from Ben Martin's version of 'dark innovation'. For him, dark innovation is a result of deficits in technique: 'the challenge to the next generation of researchers is to conceptualize, define, and come up with improved methods for measuring, analysing and understanding "dark innovation"' (Martin, 2016, p 434). But my goal is not to produce a set of 'less biased' methods that yield 'more objective' observations. Different methods only bring forward different understandings. Instead, I reframe 'dark innovation' as a call to deconstruct the central assumptions of innovation studies. Following the work of John Law, my goal is to 'scrape away the self-evident to understand and question how methods structure the world' (Law, 2016, p 53). This book is not about filling knowledge gaps through improved methods; it is about problematizing the decisions that create dark innovation in the first place – the implicit and explicit choices researchers make about what 'counts' as innovation.

Problematization

Of course, some understudied forms of innovation are already being advanced through a gap-finding logic. I could have done the same in this book. Starting from some hint in the literature or a personal hunch, I could have searched for and attempted to observe some previously unreported innovation phenomena. I could have asked: 'Where might we observe dark innovation?' (Or, more directly, I could have asked something like: 'Do public organizations produce innovative goods?') However, that kind of gap-spotting logic would only yield an incremental contribution (Alvesson and Sandberg, 2011; Sandberg and Alvesson, 2011). The grand challenge would remain untouched. This is because gap-finding maintains a high degree of path dependence in any research field (Palmer, 2006). As Alvesson and Sandberg state, 'gap-spotting is more likely to reinforce or moderately revise, rather than challenge, already influential theories' (2011, p 25). Yet, it is the most common approach to framing research questions in the social sciences (Alvesson and Sandberg, 2011). It is ironic that this gap-spotting norm has taken hold in innovation studies – where most researchers applaud novelty. If we aspire towards radical and disruptive insights, we need problematizing research questions (Sandberg and Alvesson, 2011). We need to identify, question, and undermine our social scientific norms. And so, this book asks: why do we struggle to observe dark innovation?

We already know that novel contributions to innovation studies (IS) are experiencing a 'rough ride' through peer review because of overly strict adherence to disciplinary conventions (Martin, 2016, p 440). Leading innovation researchers have warned that the field is beginning to struggle with 'disciplinary sclerosis' – the rigidity that comes from standardizing and normalizing as a discipline (Fagerberg et al, 2013; Martin, 2013). Innovation researchers have been closing ranks around certain theories, research questions, empirical contexts, and methods. Within the agreed boundaries of 'the field', there is disciplinary enforcement of shared values and practices. There is the expectation of a 'fairly standard form' of academic writing (Martin, 2016, p 440). The field of innovation studies is paradigmatically stuck.

In response, Jan Fagerberg has argued that it would be better to accept innovation studies as an interdisciplinary 'mongrel' (Fagerberg et al, 2013, p 11). Apparently, this means abandoning aspirations towards scientific 'pedigree' and instead 'engaging in fruitful intercourse' with neighbouring fields, like business history and science and technology studies (STS) (Martin, 2016, p 440). Even setting aside the elitist metaphor about breeding, this suggestion is naïve. We are talking about fields with drastically different social scientific norms – different 'epistemic cultures' (Knorr Cetina, 1999). They have not always been so different. In the 1960s and 1970s, notable academics like John Langrish, Donald Marquis, Nathan Rosenberg, Michael Gibbons, and Michel Callon moved between the subjects of innovation, science, and technology. But one of the few recent instances where innovation studies has accepted 'inputs' from STS has been the area of innovation for sustainability, and that work 'was regarded as rather "flaky" by some in IS' (Martin, 2016, p 435). Culture clash is now inevitable; in fact, I think we should encourage it.

In this book, I use ideas from elsewhere to show that dark innovation is a byproduct of innovation studies norms. I can do this because I am a bit of a scholarly mongrel. During my PhD studies, I learned that my values and assumptions do not fit within mainstream innovation studies. That PhD research became the foundation for this book. But this work does not fit any better into any other field. I am not quite a business historian, although I have training and plenty of colleagues in that area. I have no claim to affiliation with STS, although I have read a great deal of STS work. Nor am I a geographer, although my master's degree was in that field. I am too critical to be a 'proper' business or management scholar. And as I learned in the review process for this book, I am not sufficiently concerned with *labour* to fit neatly into critical management studies (CMS). In the end, I am not concerned with fitting into a discipline or field; instead, I am concerned with disrupting disciplinary convention. In the following chapters, I will use instrumentalities from STS, CMS, critical organizational history, and critical geography to confront the rigidities of innovation studies.

Instrumentalities

The empirical material for this book comes from an area of science and technology where disciplinary boundaries are very fluid (pardon the pun). I examine innovation in ocean science instruments. In Chapter 6, it will become abundantly clear that ocean science – sometimes called oceanography – is 'not so much a science as a collection of scientists' (Bascom, 1988, p xiii). These scientists come from biology, physics, chemistry, mathematics, and other disciplines. Many of them have been united around their own dark matter challenge: we currently have better images of the Martian surface than 85 per cent of our ocean floor. We are 'in the dark' about much of what lies below the ocean surface. Throughout this book, I will show that we are also in the dark about the key role of public organizations in developing the necessary new tools and techniques to understand our changing ocean. These are devices like the wave-powered ocean profiler that was developed in the mid-1990s at the Government of Canada's Bedford Institute of Oceanography (BIO) (Fowler, 1997). That 'instrument assembly' (Fowler, 1997, p 1) offers an energy-efficient way to collect time-series observations of ocean conditions. It 'utilizes ocean wave energy to provide power for repeated ascent and descent' of a sensor assembly (Fowler, 1997, p 1). Those who would like to 'geek out' on the details can read US Patent 5,644,077 (Fowler, 1997) or the plainer language description provided to the *Marine Technology Society* by Fowler et al (1997). I will share brief examples of other such devices throughout this book: salinometers, hydrophones, variable depth sonar systems, acoustic tracking tags, photosynthetically active radiation sensors, and the like. But while I draw attention to the presence of these gizmos and gadgets, I do not want to get carried away by the details of how they operate. Instead, I want to get carried away by the details of the tools and techniques that help us (fail to) observe such innovations.

Despite their importance, we often fail to notice new scientific instruments and techniques. Nathan Rosenberg once said that 'the emergence and diffusion of new technologies of instrumentation ... are central and neglected consequences of university basic research' (1992, p 381). Indeed, instruments and techniques likely constitute 'much of the "technological output" of the university system' (Salter and Martin, 2001, p 523). Later, we will see that 'output' is a crass, linear simplification. But still, 'surveys of the relationship between science and industry tend not to consider the role of instrumentation and methodologies in any detail and to discount their importance' (Martin et al, 1996, p 22). Ammon Salter and Ben Martin have argued that this is 'because of the limited ability of industrial R&D managers to recognize the contributions made by earlier government-funded research' (2001, p 522). Alternatively, Bernward Joerges and Terry Shinn suggest that 'since it [research technology] is very much a phenomenon 'in-between' and relatively invisible to outside observers, it is not surprising that it has gone largely unnoticed by students of science and technology' (2001, p 11). And Peter Galison has observed that even within an instrumentation-heavy science like physics, the instruments are easily disregarded as merely 'engine grease' that enables the more interesting 'experimental results and theoretical constructions' (1997, p xvii). Whatever the rationale, scientific instruments are underestimated and understated.

However, as I have already noted, some of the most important contributions to innovation studies quietly arose from the study of scientific instruments. There might be little acknowledgement of these technologies in innovation studies, but researchers in the history and philosophy of science have given considerable thought to questions of scientific instrumentation (see de Solla Price, 1984; Galison, 1997; Joerges and Shinn, 2001; Baird, 2004; Taub,

2011; Marcacci, 2019). Perhaps the most influential of these was Derek de Solla Price. Submitted only months before his sudden death, de Solla Price's (1984) paper on 'the science/technology relationship' provides a sweeping description of innovation in scientific instrumentalities. Drawing on examples from history, de Solla Price outlined the critical relationships among those who perform science and those who craft scientific techniques. For him, it was important to separate the products of science from the processes of science. The products, or outputs, of science can be described as scientific *know-what*. Some ancient Greek philosophers called these knowledge products *episteme* – that is, understandings and beliefs. They used the term *techne* for the *know-how* of science – that is, the processes or craft. In lieu of the Greek term, de Solla Price (1984, p 3) wrote about scientific 'instrumentalities'. Here, he aimed to capture both instruments and techniques. He said that an instrumentality could be any 'laboratory method for doing something to nature or to the data at hand' (de Solla Price, 1984, p 13).

I would not confine instrumentalities to laboratories, but I agree with de Solla Price's (1984) broad and pragmatic definition. I am happy to use the term 'scientific instrumentalities' for any materials or techniques that anyone claims to use for scientific ends. This helps me consider both the physical devices of ocean science and the less physical techniques or 'methods' deployed in innovation research. In a roundabout way, the word 'instrumentality' blurs the problematic distinction between goods and services – the tangible and tacit. More importantly, de Solla Price's approach avoids closure around the question of 'what is a scientific instrument?' (Warner, 1990; Taub, 2019).

Curators of science museums have been especially concerned with this question (Warner, 1990; Taub, 2019). Deborah Warner (1990) of the Smithsonian once pointed out that the term 'scientific instrument' only developed in the 19th century and it has always been contested. More recently, Liba Taub has shown that 'there has not been (and is not) always one universally agreed answer to the question "what is a scientific instrument"' (2019, p 454). In her study of how scientific instruments have been defined over time, Taub identifies a point at which the *Oxford English Dictionary* began to make a classist distinction between the words 'tool' and 'instrument'. The former became associated with 'workman or artisan' and the latter with 'more delicate work or for artistic or scientific purpose' (*Oxford English Dictionary*, cited in Taub, 2019, p 455). She argues that the label 'instrument' thereby came to signify professional or disciplinary status. And so she advises that museum curators 'need to ask who defined 'scientific instruments', why and how?' (p 453). Here the emphasis is on physical artefacts, but the same questions should be asked of all scientific instrumentalities.

Across the social sciences, we know that certain instrumentalities have prestige, especially standardized survey 'instruments'. You are more likely to be considered a 'proper' social scientist if you use the 'proper' methods. Of course, these preferred instrumentalities vary by discipline. Standardized scales and randomized experiments are the norm for some fields. Elsewhere, fieldwork is preferred. In some disciplines, numbers carry all the weight. In other fields, ethnographic storytelling is the preferred way of knowing. This is how scientific instrumentalities make up part of what Karin Knorr Cetina called 'epistemic cultures': 'those amalgams of arrangements and mechanisms – bonded through affinity, necessity, and historical coincidence – which, in a given field, make up *how we know what we know*' (Knorr Cetina, 1999, p 1, emphasis in original). Scientific instrumentalities are cultural artefacts and cultural practices.

This idea is reasonably well studied in the physical sciences where there are a great many physical instruments. For example, Peter Galison's history of instrumentation in microphysics can be read as a set of stories about 'changing values and meanings as they are read into and out of the knowledge machines we call instruments' (1997, p 63). This perspective treats scientific instrumentalities as more than methodological algorithms. It sees them as 'encultured' or 'entangled' (Galison, 1997, p 4) within a 'complicated patchwork' (Galison, 1997, p xx) of scientific practices. The instruments become worthy of our attention 'if they are understood as dense with meaning, not only laden with their direct functions, but also embodying strategies of demonstration, work relationships in the laboratory, and material and symbolic connections to the outside cultures in which these machines have roots' (Galison, 1997, p 2). In this book, I extend this understanding to the instrumentalities of innovation studies. There are fewer noticeable physical instruments in this field. So, while Galison writes about instruments as 'material culture of a discipline' (1997, p 2), I am writing about instrumentalities as key components in the *sociomaterial* construction of a discipline.

It is not that innovation studies have unique instrumentalities, like the particle colliders of high-energy physics or wave-powered profilers of oceanography. Rather, it is that disciplinary cohesion has emerged around a shared interest in public policy – which has long been tied to instruments of quantification (Alonso and Starr, 1987; Rose, 1991). And so, if there is a coherent 'innovation studies' discipline, then its cultural roots are planted in the numbered and neoliberalized soil of public policy. This is not to say that a straight line can be drawn between the policies of Margaret Thatcher or Ronald Reagan and present-day innovation research. It is rather more like how Cosmo Howard describes the neoliberalization of Australian and Canadian official statistics: 'a complex and only partially coherent assemblage of calculative rationalities, technologies, and practices' (2016, p 132). This book will confront a variety of scholarly practices that carry neoliberal assumptions – practices that neoliberalize ideas about innovation.

This means that instrumentalities are not merely technical – they also have an epistemic nature. They make knowledge and they contain knowledge. This has been a recurring argument in the philosophy of science since the time of de Solla Price. In his book on scientific instruments, Davis Baird noted that 'a major epistemological event of the mid twentieth century has been the recognition by the scientific community of the centrality of instruments to the epistemological project of technology and science' (2004, p 19). He argues that 'instruments need to be understood epistemologically on par with theory' and that 'instruments themselves express knowledge of the world' (2004, p 31). In so doing, he develops a way of thinking about the 'thing knowledge' embodied by scientific instruments. I will visit Baird's ideas in Chapter 2. For now, it is enough to recognize that scientific instrumentalities sit within a 'fluid relationship' (Baird, 2004, p 32) between science (*episteme*) and technology (*techne*). In other words, scientific instrumentalities are part of what Bruno Latour called 'technoscience' (1987, p 29).

Before I go too much further, it is also important to notice the fluid mobility of instrumentalities beyond their original uses. Joerges and Shinn (2001) address this in their book, *Instrumentation between Science, State and Industry*. They use the label 'research technologies' for one subset of scientific instrumentalities:

> instances where research activities are orientated primarily toward technologies which facilitate both the production of scientific knowledge and the production of other goods. In particular, we use the term for instances where instruments and methods traverse numerous geographic and institutional boundaries; that is, fields distinctly different and distant from the instruments' and methods' initial focus. (Joerges and Shinn, 2001, p 3)

From this perspective, many scientific instrumentalities have 'interstitiality' or 'trans-community positioning' (Joerges and Shinn, 2001, p 7). They 'link universities, industry, public and private research or metrology establishments, instrument-making firms, consulting companies, the military, and metrological agencies' (Joerges and Shinn, 2001, p 3). This means that advancements in research technology can have far-reaching effects on science, industry and government (Joerges and Shinn, 2001).

As we will see in Chapter 2, the theoretical models of innovation studies have struggled to account for this. Scientific instrument innovation has been observed through linear push, market pull, chain-link, and system models. However, all these 'reductive schemes' (Galison, 1997, p 15) have been rejected by close studies of scientific instrumentality innovation that used post-positivist methods (see Galison, 1997, p 15; Joerges and Shinn, 2001, p 4). Galison puts it plainly when he says, 'the dispersion of instrument knowledge follows no univocal pattern' (1997, p xxi). For this reason, scientific instrumentalities are especially helpful empirical materials for disrupting the patterns of innovation studies.

Overview of methods and chapters

Throughout this book, I use ocean science instrumentality innovation as an empirical motif for exploring the instrumental biases in innovation research. This is a double entendre. I am writing about the need for innovation in social science instrumentalities. I am also writing about innovation in ocean science instruments. I examine archival records and analyse structured interview data from a regional concentration of ocean science and technology organizations where I live on Canada's Atlantic coast (see Figure 1). Using a variety of post-positivist methods from outside innovation studies, I show how public organizations have developed novel technological goods, while interacting symbiotically with private companies that are 'quartermasters' for this scientific enterprise.

It will become clear that my work was originally conceived as a snapshot study – albeit one with substantial historical background material. When I presented this work as a monograph PhD thesis, I focused on the question of public innovation in goods and strayed only slightly from innovation studies norms. The 'messiness' (Law, 2004) was constrained. But as I have said, this book is concerned with how disciplinary norms constrain the mess. To give this work an 'after method' (Law, 2004) sensibility, I reframed it in the spirit of the 'biographies of artifacts and practices' (BOAP) approach from STS (Hyysalo et al, 2019). Sampsa Hyysalo, Neil Pollock, and Robin Williams recently argued that 'if STS is to continue to provide insight around innovation this will require a reconceptualization of research design, to move from simple "snap shot" studies to the linking together of a string of studies' (Hyysalo et al, 2019, p 4). I agree. Across the next seven chapters, I do as they suggest and 'knit together different kinds of evidence – that includes historical studies, ethnographic research, qualitative studies of local, and broader development' (Hyysalo et al, 2019, p 16). I make inquiries into my theoretical and contextual points of departure. I question different technological framings by allowing 'ocean science instrumentalities' to float within the categories of 'scientific instrument' and 'ocean technology'. Most importantly, I play with methods, because these are the real research subjects for this book.

Law challenged us 'to imagine what research methods might be if they were adapted to a world that included and knew itself as tide, flux, and general unpredictability' (2004, p 7). And as we will see, this is the world of dark innovation. The pages that follow include wave after wave of concepts, theories, technologies, research areas, references, and writers. There will be

Figure 1: Map of Canada's Atlantic coast (including major cities in the US for reference)

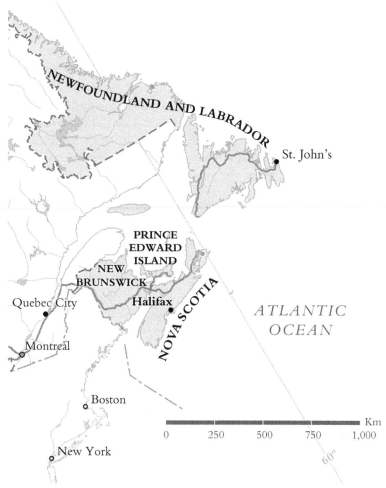

Source: Author's adaptation from the Atlas of Canada (*Canada with Names*, 2022) under the Government of Canada's Open Government License.

calm moments, but I will also be asking a lot of readers. By the end of this book, we must have at least breached the instrumental 'sea wall' that has been keeping innovation studies dry.

In Chapters 2, 3 and 4, I will show how critical organizational historiography (Wadhwani and Bucheli, 2014; Durepos et al, 2021) can help reveal dark innovation. Schumpeter once argued that many of the important questions about innovation are best tackled with the tools of history rather than statistics (Godin, 2017, p 64). Over seventy years later, research on innovation remains dominated by econometrics. It tends towards the 'don't ask, we won't tell' (Gaddis, 2004, p 92) stance on historical methodology that permeates organization studies (Wadhwani and Bucheli, 2014).

First, in Chapter 2, I build on the excellent genealogies of innovation theory developed by Benoît Godin (2006, 2017, 2011, 2012; Godin and Lane, 2013). I follow those working in the historiography of management knowledge (such as Cooke, 1999; Dye et al, 2005; Kelley et al, 2006; Genoe McLaren et al, 2009) and argue that literature reviews are scholarly histories. Here, I am interested in showing how some ideas are excluded from our sense of the theoretical past. Since the extant literature is our epistemic point of departure, our research is always cognitively constrained by our sense of the past. I will show that scientific instruments are important missing characters in the stories we tell about major innovation theories. When those theories are cited and discussed today, they are black boxed in such a way that the scientific instruments are easily forgotten. In response, Chapter 2 is a history of innovation theory that centres scientific instrument innovation. I consider why scientific instruments might be missing from most stories of past innovation research and I quote personal correspondence with Eric von Hippel on how his colleagues were initially dismissive of scientific instruments as an overly 'abnormal' field of study. Thus begins my exploration of some of the scholarly processes that produce dark innovation, and some of the instrumentalities that can uncover hidden or marginalized subject matter.

In Chapter 3, I problematize the taken-for-granted nature of 'context' and demonstrate an alternative way of 'practicing context' (McLaren and Durepos, 2019, p 74). It is quite normal for research in innovation studies to include some discussion about the history of a regional-industrial context before engaging with primary data. But readers are typically asked to take the authors' expert knowledge of the historical context for granted. Instead, I use an ANTi-History approach (Durepos and Mills, 2012) to (re)assemble three histories of one ocean science and technology sector in Nova Scotia, Canada. I present three incompatible newspaper and magazine accounts of this sectors' emergence – from 1960, 1980, and 2012. The earliest account of the sectoral history positions scientists, scientific instruments, science organizations, and geopolitics as key actors. But in the latest account, scientific instruments are not present; the main actors are private companies and science is lauded as the knowledge base that spawned these companies. I argue that these three different histories are traces of efforts to define a sector/cluster/industry identity and to rhetorically impose that identity on various actors. I argue that these are 'rhetorical histories' (Suddaby et al, 2010) that aim to 'assemble' a cluster as historical fact, thereby establishing a regional competitive advantage. By treating the industrial history in this way, I demonstrate the need to take historical method (that is, historiography) seriously in research on innovation.

In Chapter 4, I consider the narrative tools that shape our understandings of innovation. The apolitical treatment of science and technology (Fagerberg et al, 2012b; Martin et al, 2012) – including the apolitical treatment of the state (Pfotenhauer and Juhl, 2017) – has already been flagged as an issue in innovation studies. I advance those discussions by directly surfacing the political nature of ocean science and technology in Nova Scotia. I argue that when we turn to the past from a present-day neoliberal standpoint, we always end up writing stories about market-dominant evolutionary processes. In contrast, this chapter presents the stories of three public research organizations (PROs) and the politics around their establishment. In these organizational settings, I find private industry becoming enrolled in political missions of military defence, Canadian sovereignty, and scientific one-upmanship. Analysing these short histories as a set, I find that the public organizations can be characterized as active political agents. Meanwhile, the private companies around them can be characterized as 'quartermasters' – like the individuals responsible for providing supplies to units in an army (or the Q Branch in *James Bond*). They were producing the scientific instrumentalities needed for multiple 'cold wars'. This relationship was more nuanced than the simple provisioning of equipment and services – it was often a close two-way partnership. The technical expertise provided by scientific instrument companies helps to set the course for science, and vice versa. Telling the past in this way makes the boundary between public and private organizations messier than it appears in neoliberal ideology.

At the halfway point of this book, I shift focus from historiographic methods to modern day acts of counting and classification. First, in Chapter 5, I explore the puzzle of innovation taxonomies. Anecdotally, I consider how strangely normal it is for people in Atlantic Canada to speak of an 'ocean technology sector'. Does these mean there are only two other types of technology: land and aerospace? Ocean technology is very clearly a 'folk taxonomy'. It is akin to the way in which many people speak of spiders as 'bugs'. This makes it a useful category for demonstrating that sectoral boundaries are not as 'natural' as innovation research assumes.

Product-based industrial classifications (Standard Industrial Classification, North American Industrial Classification System, etc.) are embedded within innovation theory through the methods that were used by Keith Pavitt (1984) and his heirs (Archibugi, 2001; Castellacci, 2008). The resulting 'taxonomies' of innovation have been widely used in research and public policy (de Jong and Marsili, 2006). I argue that all this activity is driven by one of the many biological metaphors that permeate innovation studies. The idea of taxonomic classification is one of those that we have explicitly borrowed from biology (see Archibugi, 2001; de Jong and Marsili, 2006). I use analogies from taxonomic biology to reconsider three major methodological problems already described in the innovation taxonomies literature. This includes, but is not limited to, the taxonomic separation of public and private organizations. I then turn to the deeper problems that arise from the implicit use of an organizations-as-organisms analogy. Following the work of Gareth Morgan (1980, 1986), I consider the metatheoretical implications of biological metaphors and the corresponding political assumptions that are inscribed in sectoral classification tools. I argue that we will need other metaphors if we hope to observe other innovation patterns.

I take the question of metaphors one step further in Chapter 6. There, I approach 'region' as a spatial metaphor that both enables and constrains the systems of innovation literature. Thinking about innovation as a regional phenomenon allows for the surveyable, measurable, Euclidean spaces we call innovation systems. But this makes it difficult to observe innovation processes that fold people, things, and places together in new ways. Taking 'object lessons' from John Law and Vickey Singleton (2005), I explore the different possibilities that come from framing innovation as a region, network, fire, or fluid object. First, I unpack the regional metaphor by describing the boundary choices I made while preparing a survey of ocean science instrumentality innovation. Then, I share 'excess' network data that extended beyond my region and other observations that would have 'flooded' or 'set fire' to the boundaries I had established. On these grounds, I question the 'hegemony' of regionalism (Sepp, 2012, p 47) within innovation studies. Many innovation scholars would say that we need the regional metaphor so we can pin things down, survey them, and quantify them – otherwise they do not count. But I suggest a turn towards critical geography, where other topological metaphors provide ways 'of understanding space and time when the numbers no longer quite add up to anything significant' (Allen, 2011a, p 316).

In Chapter 7, I dig deeper into the idea that good numbers can be meaningless. We know that there are problems embedded in the construction of standardized innovation measures (see Godin, 2002, 2005; Gault, 2018; Gault, 2020). To address some other problems of statistical practice, I develop and deploy *autoethnostatistics*. This is a fusion of autoethnography (Ellis, 2004) and ethnostatistics (Gephart, 1988, 1997, 2006). Autoethnography is cultural research that uses narrative inquiry into one's personal experiences (see Ellis, 2004; Prasad, 2019). It has seen limited use in innovation research (for example, Rehn, 2023), but it is a well-known approach elsewhere. Meanwhile, ethnostatistics is 'the empirical study of how professional scholars construct and use statistics and numerals in scholarly research' (Gephart, 2006, p 417). It is generally underutilized (Gephart, 2006; Helms Mills et al, 2006) and has also seen limited use in innovation research (see Kilduff and Oh, 2006). Rather than producing a distanced critique of how *other people* do statistics, I combine these methods to examine my own experiences inside the cult of numbers that dominates mainstream innovation studies.

My work in Chapter 7 might challenge both critical and mainstream scholars. Following the norms for reporting statistical work, I present four analyses of my data on ocean science instrumentality innovation in Nova Scotia. From the outset, simple descriptive statistics provide evidence to falsify any claim against the existence of public innovation in goods. But these numbers also lack any measure of statistical significance. And so, I follow the descriptive statistics with three sets of statistically significant results. These results follow the canonical progression of innovation theory discussed (and debunked) in Chapter 2. After each analysis, I break into an autoethnographic discussion on the meaning(lessness) of those results. Using Albert Camus' *The Myth of Sisyphus* (1955), I build directly upon autoethnography's 'existential convictions' (Bochner, 2012, p 209). I suggest that following convention is like being condemned by the statistical gods to push numbers up a hill, hoping to successfully reach the summit, only to realize that the effort was meaningless and alienating.

In the final chapter, some readers might expect me to lay out a research agenda for finding dark innovation. I do suggest some opportunities arising from this book. But my focus is on how we might further deconstruct the instrumentalities of innovation research. To this end, I deconstruct some of my own instrumentalities. I finally notice the most influential oceans-related innovation to ever emerge from my part of the world and I consider how the 'epistemic culture' of innovation studies pointed me and my tools away from it. Returning to the dark matter analogy, and with insights from both Karin Knorr Cetina (1999) and Karen Barad (2007), I explain why some physicists might think their observations are meaningless and why this leads them to obsess over their instrumentalities. I argue that we need a similar humility and obsession in research on innovation.

2

Extant Theory

Joseph Schumpeter is widely regarded as the father of innovation theory (Fagerberg et al, 2012a; Lundvall, 2013b). Although other economists began citing his work in the 1940s and 1950s (Godin, 2020), widespread engagement with his ideas about entrepreneurship and innovation did not occur until the 1980s – more than 30 years after his death. First, there was the lag in translation of his early work. Until he moved to Harvard in 1927, Schumpeter wrote in German. A full 20 years passed before many of his major works were available in English (see Schumpeter, 1934, translated by R. Opie). Even then, his *The Theory of Economic Development* (Schumpeter, 1934) lost a full chapter between its 1911 original German printing, a 1926 German reprint, and the 1934 English translation. More remarkably, the concept of innovation was *added* in this process (Godin, 2019). The 1911 original dealt with combination, not innovation (Godin, 2019). Indeed, much was lost or changed in the translation/transition of Schumpeter's overall oeuvre. Jürgen Backhaus explains that, despite Schumpeter's bilingualism, 'when he writes in English he has to cast the argument differently in accordance with the different writing style, but also intellectual tradition' (2003, p 1). And so, beginning in the earliest days of innovation scholarship, the field has been strewn with lost, forgotten, and 'translated' bits of theory.

The late Benoît Godin has been the only serious historian of innovation. He asked the following question: how did innovation come to be known as it is today? Godin (2020) longed for innovation policy to have many more historians, like science policy. Indeed, there are rich and relevant academic traditions in the history of science (for example, Kuhn, 1962) and the history of technology (for example, Bijker et al, 1987). And there is burgeoning interest in entrepreneurial history (for example, Cassis and Minoglou, 2005; Landström and Lohrke, 2010; Wadhwani and Lubinski, 2017). But history is still crowded out of innovation studies by econometrics. This is a problem because 'many economists are bad historians, or simply not historians at all' (Godin, 2017, p 76). Godin tackled this deficit across his many books and articles. He examined the history of innovation models (Godin, 2006,

2017, 2011; Godin and Lane, 2013) and measurements (Godin, 2002, 2005), the rise of ideas about technological innovation (Godin, 2019, 2020), and the construction of innovation studies as a disciplinary field (Godin, 2012, 2014; Godin and Vinck, 2017a). In this chapter, I take a similarly genealogical approach to ideas about innovation. I use historiography as a tool for reconsidering extant innovation theory – where 'extant' literally means 'still in existence; surviving' and comes from the Latin for 'being visible or prominent' (*Oxford English Dictionary*). My aim is to make visible some marginalized but *material* aspects of innovation theory.

However, this chapter is not a 'corrected' and comprehensive history of knowledge about innovation. Rather, my focus is on one of many elisions. I am interested in how scientific instruments went missing from the stories we tell about innovation theory. As I noted in Chapter 1, scientific instruments were key actors in the theory development of Eric von Hippel (1976, 1986), Christopher Freeman (1974), Stephen Kline (1985; Kline and Rosenberg, 1986), and Nathan Rosenberg (Rosenberg, 1982; Kline and Rosenberg, 1986). Through these men, scientific instruments were central to four of the top ten most-cited works in innovation studies (Fagerberg et al, 2012a). But when these theories and models are cited and discussed today, they are black boxed in such a way that the scientific instruments are forgotten. In response, this chapter is a history of innovation theory that centres on scientific instrument innovation. This is similar to the way that Vinsel and Russell (2020) foreground maintenance in a chapter on the history of technological innovation. But while their goal is to revalue maintenance, my goal is a different understanding of innovation models.

My discussion will follow the commonly accepted junctures and chronology of innovation studies – I begin with linear models in the aftermath of the Second World War and proceed through to the innovation systems approach. We will see that empirical research on scientific instruments shaped the linear model debates, the chain-linked model, and the innovation systems approach. But over time, the influence of instrumentation research becomes muted. Models of innovation start to position scientific research as a support function for technological development in private business. Scientific instruments are written off as 'unusual' because they do not fit this mould. By the end of this timeline, innovation research has turned towards 'normal' market-based technologies.

There is a potential drawback to plotting this review along a timeline. This chapter might look like a progressive and teleological account – one that evolves towards the dominant present-day model of the 'innovation system'. But my intent is not to suggest that 'old' ideas have disappeared. Readers should not assume that ideas falling early on my timeline are now gone. For example, the linear model is still evident today (Godin, 2017). And, as we will soon see, readers should not assume that the innovation systems approach is dominant today because it is a newer idea than the others. Indeed, the theoretical contributions I discuss in this chapter are connected through a common technological context and time period. In Chapter 7, I will put linear, chain-linked, and systems models side by side to consider their present-day analytical liveliness. But for now, let's go back to the 1940s and consider how scientific instruments joined the entanglement of ideas regarding science, technology, and innovation in the US.

Instrumental innovation models

'Linear' developments

There was certainly theorizing about innovation before the Second World War. There were many sociologists, anthropologists, economists, business school professors, and industrialists proposing stage models of cultural and technological change from the 1920s to the 1940s (Godin, 2017) (this was before any anglophone had a chance to read Schumpeter). Indeed, Godin (2017, 2011) credits Maurice Holland – a director at the US National Research Council – with articulating the core ideas of the linear innovation model during the 1920s. But it was certainly the success of science during the Second World War – particularly the Manhattan Project – that solidified linear thinking within a 'post-war paradigm' of popular and academic theorizing (Nemet, 2009).

The most prominent voice for this paradigm was US presidential advisor Vannevar Bush. He had led the wartime US Office of Scientific Research and Development. And it was his report – *Science, the Endless Frontier* (Bush, 1945) – that most famously argued for postwar public investments in science to radically advance medicine, industry, and national defence (1945). There is no doubt that the Bush report was widely read and influential. It is frequently cited as either the source or turning point towards the linear model of innovation (for example, Irvine and Martin, 1984b; Freeman, 1996; Lundvall, 2013b). However, Bush did not elaborate an innovation model – sequential or otherwise (Godin, 2017). His report merely argued for a causal link between basic research and socioeconomic progress. Godin (2017) suggests that the closest Bush came to articulating a linear model was through his connection to Rupert Maclaurin.

Maclaurin, an economic historian at the Massachusetts Institute of Technology (MIT), had been secretary on one of the four committees that contributed to Bush's report. He had also been a student of Schumpeter, whose ideas he later developed into a staged innovation process (Maclaurin, 1949, 1950) – a theoretical framework that would eventually become known as the linear model (Godin, 2008, 2017). Along the way, however, Maclaurin's work would be forgotten; qualitative research was not respectable in economics at that time (Godin, 2008, 2017). We would also forget the many others, over many decades, who developed various stage and technology-push models. These would be subsumed under the 'linear model' label around the 1960s. So, despite the canon, Bush did not give us the linear model.

But Bush did influence the thought paradigm in which the linear model solidified: making science instrumental to technological change. In other words, science became a tool for producing technology. Pfotenhauer and Juhl go so far as to say that Bush's report 'castrates the government mandate by confining the state's responsibility to the front end of the pipeline' (Pfotenhauer and Juhl, 2017, p 72). This fundamentally reframed the role of government with respect to science, technology, and innovation. The idea that advancements in basic science drove technological progress then 'held sway for 20 years or so' (Martin, 2010, p 3). And, as we will see in Chapter 7, it continues to lurk in the background today through standardized innovation statistics (Godin, 2017).

Along the way, a debate emerged about the direction of the relationship between technoscientific progress and market demand. Those who ascribed to the 1940s postwar paradigm described an innovation process that was initiated and driven – or 'pushed' – by advancements in technoscientific knowledge. But in the 1960s, an alternative hypothesis emerged suggesting that market demand served to 'pull' – or determine the speed and direction of – innovation (Schmookler, 1966; Rosenberg, 1969). These two positions were entrenched by results from the US Department of Defense's HINDSIGHT project in 1966 (arguing for 'pull') and the US National Science Foundation's TRACES project in 1969 (arguing for 'push'). In retrospect, we reject both positions as highly simplistic. Although he was a key proponent of demand-pull, John Langrish (Langrish et al, 1972; Langrish, 1974) now characterizes the whole debate as somewhat silly (Langrish, 2017). In describing it, he invokes the imaginary two-headed 'pushmi-pullyu' creature from the *Dr Doolittle* books (Lofting, 1920). Most stories about innovation theory gloss over the details of this silliness, especially the short-lived arguments for a demand-pull model (Godin, 2017). But Godin is right that this was a critical juncture in the shift towards a market bias for innovation research.

Under the postwar paradigm, there had been considerable interest in meeting military and social needs (for example, healthcare). 'Those needs concerned public decisions made in the 'national interest' and had nothing to do with the "free market"' (Godin, 2017, p 121). But 'needs' were entirely subsumed by 'demand' once Sumner Myers and Donald Marquis published their frequently cited demand-pull model (Myers and Marquis, 1969). In his detailed discussion of this period, Godin explains:

> One reason for the shift from need to demand in the vocabulary and
> related analyses is that scholars chose to study technological innovation

in the context of the firm and related market factors. As the title of most studies on technological innovation attest (from Sumner Myers and Donald Marquis onward), researchers focus on firms as originators of innovation and their environment rather than public organizations as sponsors or societal needs ... When the nonmarket environment (such as government) is considered, it is studied as a market (the demand from government or government as a purchaser of new products) – or as a barrier to industrial innovation. (Godin, 2017, p 120)

So, while the demand–pull model was quickly rejected, it had a lasting influence on the language of innovation studies. The push versus pull debate also triggered an influential series of studies on scientific instrument innovation (Utterback, 1971b, 1974; Freeman, 1974; von Hippel, 1976, 1988; Rosenberg, 1982).

'Pushmi-pullyu' devices

In the late 1960s and early 1970s, scientific instruments became a key empirical testing ground for the demand-pull model. As Chris Freeman and Luc Soete would later explain, 'the increasingly intimate relationship between new materials, new process development and fundamental research is nowhere more apparent than in the field of instrumentation' (1997, p 128). In other words, scientific instruments were as close as anyone could come to studying a real pushmi-pullyu – it was not immediately clear which end was the front. And, unsurprisingly, studying them eventually moved the debate beyond one-way linearity.

Daniel Shimshoni's PhD thesis at Harvard (1966)[1] may have been the earliest study focused on innovation in scientific instruments. Trained as an engineer at Princeton and CalTech, Shimshoni helped build bombers during the war. He then led the development of the Israeli Air Force and became the first Director of the Israeli National Council for Research and Development. Returning briefly to the US, he completed his doctoral thesis on the interorganizational mobility of scientist-entrepreneurs in the instrument industry (Shimshoni, 1966, 1970). In his thesis (1966) and a subsequent paper in *Minerva* (1970), Shimshoni concluded that

an overwhelming majority of instrument innovations involved the movement of technical leaders to form their own companies or to join recently established firms. The scientific basis and the essential enabling technology of most of the innovations considered in the present study originated in university, government or large industrial laboratories, while new instrument products were largely the work of small firms. (Shimshoni, 1970, p 85)

And so, Shimshoni's study came down on the side of a push model – albeit with people moving in all directions. He therefore argued for policy makers to emphasize public research and development (R&D) expenditures and to encourage employment mobility for all those involved. He then turned to a scholarly focus on Israeli politics (including science and technology policy). However, his scientific instrument study would have a direct, cited influence on some of the most famous innovation scholars, including James Utterback (1969, 1971a, 1971b, 1974), Eric von Hippel (1975, 1976, 1988), Nathan Rosenberg (Mowery and Rosenberg, 1979), and Chris Freeman (1974, 1982; Freeman and Soete, 1997).

Only a few years after Shimshoni, Utterback would complete his own PhD on innovation in scientific instruments (Utterback, 1969). Under the supervision of Donald Marquis at MIT, Utterback set out to understand idea generation and problem solving as the 'first phases' of innovation. He examined 32 cases of scientific instrument development in the Boston area. Like Shimshoni, he recognized that spin-off companies were the main mode of entry in the instrument industry. But, unsurprisingly, his focus was on an early version of the Myers and Marquis (1969) demand-pull model. Based on his sample, Utterback concluded that new product ideas were 'predominantly (twenty-four of thirty-two cases) stimulated by information about a need' (Utterback, 1969, p 2). His eight remaining cases were 'stimulated by recognition of a technical possibility' – yet, he quickly argued, even those ideas were 'most often encountered in the course of work on a related problem' (Utterback, 1969, p 2). Utterback would go on to publish this work in *IEEE Transactions* (Utterback, 1971a), the *Academy of Management Journal* (Utterback, 1971b), and *Science* (Utterback, 1974). Hedging his bets, he would argue that 'most often' (Utterback, 1971b, p 83), given 'the weight of evidence' (Utterback, 1971a, p 131), and 'in most cases' (Utterback, 1974, p 183), 'market forces appear to be the primary influence on innovation' (Utterback, 1974, p 621). He also noted that the source of information about new instrument needs (framed as economic opportunities) was often outside the firm. Nonetheless, his study kept innovation neatly within the boundaries of the firm, 'or divisions of firms in the Boston area' (Utterback, 1969, p 2).

Two years after Utterback's journal publications, Eric von Hippel took this debate beyond the firm in his own PhD thesis (1975) and in a subsequent article in *Research Policy* (von Hippel, 1976). Here, von Hippel examined the development of 111 scientific instrument innovations in the US, including four broad classes of scientific instruments: gas chromatographs, nuclear magnetic resonance spectrometers, ultraviolet spectrophotometers, and transmission electron microscopes. He used the Myers and Marquis (1969) demand-pull model and mapped the steps undertaken by scientists – which he referred to as 'users' – versus the steps undertaken by scientific instrument manufacturing firms, which he referred to as 'producers'. He found that in nearly all cases, scientist–users performed all steps of the innovation process up to and including the development and use of precommercial prototypes. Only after this precommercial testing did producers in the private sector acquire and begin to both commercialize and incrementally improve these instruments.

Frank Spital, one of the research assistants on von Hippel's project (von Hippel, 1988, 1975), later extended the research (Spital, 1979). He added some nuance to the original observations, determining that scientist–users were responsible for major innovations and many minor improvement innovations except those that were initiated by manufacturers in response to their competitors. Nearly 20 years later, William Riggs worked with von Hippel to revisit user innovation in the context of scientific instruments. Riggs and von Hippel (1994) reconfirmed the dominant role of users in scientific instrument innovation. In their new dataset, they found 'user innovators almost never gained direct financial benefit from their instrument innovations when those were commercialized by instrument firms' (Riggs and von Hippel, 1994, p 465). They further suggested that user-driven scientific instrument innovations are more radical than producer-driven innovations, which are more incremental.

Von Hippel explained that he chose this empirical focus because Shimshoni and Utterback had already 'ascertained that innovation in response to user need was prominent in scientific instruments' (von Hippel, 1976, p 215). And so, this class of technology was ideal – not only for testing the demand-pull model, but also for building a stellar career trajectory around innovation outside the firm. Yet there was something about scientific instruments that forced von Hippel to quickly move on. He wrote that 'to explore this matter, I decided to conduct a second study in other, more "normal" fields, before suggesting that users-as-innovators might be a generally significant phenomenon' (1988, p 20). Indeed, he is now known for his work on open, distributed, and free innovation. His work is more readily associated with empirical domain of open-source software.

In preparing to write this book, I had a hunch that the line 'other more "normal" fields' was code for some resistance von Hippel might have encountered from other innovation researchers. I reached out via email and was floored by his very prompt and gracious reply, and subsequent Zoom call. With reference to that one line in the 1988 book, my email asked him: 'What was "abnormal" about scientific instruments that pushed you to confirm your findings elsewhere? Were people dismissive of your findings because of the context around scientific instruments? Was it the small market for these devices, their highly technical nature, the places these users were employed, or something else?' Von Hippel replied:

> Yes, as you surmise, that note in the 1988 book came because I needed to address the skepticism of colleagues. :-)

The problem was that my economics colleagues had a strong investment in Schumpeter's underlying assumption that innovation was done by producers.

They therefore had a strong incentive to dismiss my findings as special cases.

Most dismissed my scientific instruments findings with a comment that 'oh, that's just scientists being scientists.'

A funny additional story is that with my students, who had a strong interest in extreme sports, one of the areas I studied next was user innovation in extreme sports. My colleagues dismissed these studies also, saying in effect', everyone knows kids practicing extreme sports are crazy and not representative of anything.'

It really took the nationally representative surveys of consumers to convince my colleagues that user innovation was a general phenomenon worthy of note. (Von Hippel, 2021)[2]

This response gives us a rare insight into the processes that focus our scholarly attention one way or another. Not only do we tend to dismiss nonconformist research, but we also tend to dismiss interesting qualitative cases in favour of large statistical datasets. I will return to this latter point in Chapter 7.

For now, let me emphasize that von Hippel's (1976) paper was a turning point for innovation research. Bogers et al argue that this work was the first to notice that users can be innovators, and that it 'set off a substantial amount of research investigating users as the sources of innovation' (2010, p 859). The findings were reprinted as a key part of von Hippel's book *The Sources of Innovation* (1988), which Fagerberg et al (2012a) ranked as thirteenth on their list of top contributions to innovation studies. Over time, von Hippel's research programme helped to dispel the myth that the locus of innovation activity (von Hippel, 1976) rests within manufacturing firms. He established that the 'locus' of this activity can also rest in 'users'.

Von Hippel and his colleagues foreshadowed this decline of linear models in two ways. First, they directly observed two-way – bidirectional – interaction at the point where precommercial instruments created by users were transformed into commercial instruments by producers (von Hippel, 1976). Second, they indirectly foreshadowed the fall of this model by selecting a context where scientists were operating at both ends of the linear flow: exerting both science-push and demand-pull. Von Hippel and his colleagues positioned their language on the 'demand-pull' side of the debate. However, the 'locus' of this demand rested in those individuals who were pushing back the frontiers of science. There would have been no language available to reconcile this complexity from inside the pushmi-pullyu debate.

There would also have been no discursive space to consider the public nature of these innovations in the US in the 1960s and 1970s (which was hungover from McCarthyism). But there is no doubt that von Hippel's 'users' must have been employed in public or quasi-public organizations (for example, public or land grant universities, public research laboratories). There is no mention of public organizations in von Hippel's scientific instrument studies. There is only a passing reference in later work to 'universities' writ large (with no discussion of public/private distinctions) and 'government laboratories' (Riggs and von Hippel, 1994, pp 461–2). And yet, von Hippel's users were most definitely 'pulling' scientific instrument innovation from within the same institutions that were 'pushing' technological change. And so, let's move on from these pushmi-pullyu studies with the sense that they were unintentionally and unconsciously observing public sector innovation in goods.

'Chain-linked' processes

In the 1950s and 1960s, the linear nature of innovation was mostly taken for granted. Debate focused on directionality: technology-push and demand-pull were seen as mutually exclusive hypotheses (Chidamber and Kon, 1994; Nemet, 2009). But by the 1980s, it was widely accepted that 'innovation is neither smooth nor linear, nor often well-behaved' (Kline and Rosenberg, 1986, p 285). Research on science, technology, and innovation became loaded with words like coupling, interaction, and symbiosis (Godin, 2017). Linear models persisted, but became buried beneath layers of feedback loops (Godin, 2017).

The most well known of these layered models was Stephen Kline and Nathan Rosenberg's 'chain-linked model' of innovation (Kline, 1985; Kline and Rosenberg, 1986). It may not have been all that novel (Godin, 2017). It also had 'the problem that if you start trying to explain [the chain-linked model] to policy makers their eyes start glazing over!' (Martin, 2010, p 4). But it is often cited as the turning point away from linear models and towards a systems approach (for example, Martin, 2013). This is because the chain-linked model does not assume that innovation begins with research or with market demand. Instead, it highlights the ongoing interactions between R&D activities.

Based on his 30 years of consulting to industry, Kline proposed a 'linked-chain' model of innovation as an improvement to the 'oversimple and inadequate' linear model (Kline, 1985, p 36). In his model, Kline separated research activities (which he defined as the processes that produce knowledge) from the product development process (which he labelled as 'the chain-of-innovation') (Kline, 1985, p 36). He then argued that innovation involved not one sequential process, but five flows or pathways. In the first paper, he illustrated this across seven different figures (the paper was only ten pages long). The figures included 29 arrows indicating the flows between research, development, and stocks of knowledge. These are the details in the model that tend to make eyes gloss over. They are discussed extensively in the original papers. But for my purposes, one line is more interesting than all the others. This is the line Kline used to denote the critical role of scientific instrument innovation.

Kline labelled this the 'initiation of science link' (1985, p 41). He grounded it in Derek de Solla Price's (1984) notion of scientific instrumentalities, explaining that 'the production of new instruments, tools, and processes has in many instances made possible new forms of research' (Kline 1985, p 41). Kline's version of the paper listed the telescope, the microscope, and radiometric dating as historical examples, and this discussion was expanded in Kline and Rosenberg (1986). Kline and Rosenberg also discussed the CAT scan, the electroencephalogram, and the 'digital computer' as examples of the ongoing 'feedback *from* innovation, or more precisely from the products of innovations, *to* science' (1986, p 293, emphasis added). And so, the chain-link model recognized the important flow of new instruments and techniques into science. This flow was seen as critical to the model, but it is easily overlooked in the muddle of boxes, lines, and arrows.

However, it should be noted that scientific instruments were represented by a one-way arrow in the chain-link model. This was consistent with Nathan Rosenberg's work at the time. Kline's (1985) pathways between research and development were based in Rosenberg's earlier assertion that 'science is not entirely exogenous' (Rosenberg, 1982, p 142). In other words, the chain-link model did not consider science to be disconnected from the market; instead, it considered scientific research and technological development to be directly and indirectly linked. Rosenberg developed this sense of the links between science and markets in a 1981 conference paper. The paper appears in his book *Inside the Black Box* (1982) – another of the top twenty contributions to innovation studies (Fagerberg et al, 2012a). There, Rosenberg argued that 'improvements in instrumentation, through their differential effects upon the possibilities of observation and measurement in specific subfields of science, have long been a major determinant of scientific progress' (1982, p 158). In other words, technology pushes science. Rosenberg's understanding of scientific instrument innovation at that time appears to have been influenced by the technology-push discourse. Although the chain-link model was attempting to overcome linear flows, scientific instruments were illustrated and described as flowing in one direction. However, Rosenberg knew that this analysis was 'only the first small step on a long intellectual journey' (Rosenberg, 1982, p 142).

Indeed, Rosenberg returned to the study of scientific instruments ten years later. In a *Research Policy* paper, he noted the importance of scientific instrumentality innovation for socioeconomic development. He suggested that the primary product of basic science is knowledge about the nature of our universe, but new instrumentation techniques are an important and overlooked byproduct of this work (Rosenberg, 1992). Drawing on histories of scientific instrument innovation (including computing, magnetic resonance imaging, electron microscopy, and lasers), Rosenberg discussed the movement of instrument innovations across scientific disciplines and through various industries:

> Improved instrumentation has had consequences far beyond those that are indicated by thinking of them simply as an expanding class of devices that are useful for observation and measurement ... they have played much more pervasive, if less visible roles, which included a direct effect upon industrial capabilities, on the one hand, and the stimulation of more scientific research on the other. (Rosenberg, 1992, p 388)

Here Rosenberg echoes the recurring sentiment that scientific instrumentalities are a highly important innovation context due to their wide diffusion through society. This diffusion is at least partly thanks to the work of private industry. Like others (von Hippel, 1976; Spital, 1979; von Hippel 1988; Riggs and von Hippel, 1994), Rosenberg (1992) notes that private sector manufacturers make incremental improvements to scientific instruments. These improvements in performance, versatility, price, and usability for those with less training in the original applications of the technology help to facilitate diffusion of the innovations. But further to his collaboration with Kline (1986), Rosenberg reminds his readers that innovation is not linear. A new scientific instrumentality can stimulate follow-on research with respect to performance, materials, or ancillary technologies, as well as open new fields of research, be adapted to other fields of research, and be adapted to commercial applications (Rosenberg, 1992). He concludes that, in the context of scientific instruments, the 'scientific research community undertook radical innovative initiatives that led, in many cases, to the eventual supplying of its own internal demand and, in the process, provided large external benefits as well' (Rosenberg, 1992, p 389).

By 1992, Rosenberg agreed with de Solla Price (1984) on the widespread importance of scientific process innovations as well as the nature of the relationships between scientists and the scientific instrumentality industry. They both rejected the idea that knowledge flows one-way from science to industry via instruments or any other means. It is appropriate to think of scientific instruments as the inputs or 'capital goods of the scientific research industry' (Rosenberg 1992, p 381), yet it is also important to recognize that 'scientific instrument firms are quite often spin-offs from great national facilities in experimental science … [and] the mechanism for the entrepreneuring and expansion of such crucial high technology laboratories has been government procurement' (de Solla Price, 1984, p 18). And so, the relationship between research organizations and instrumentality companies can be described as 'interactive' (Rosenberg, 1982, p 158), 'complementary' (Rosenberg, 1992, p 386), or 'symbiotic' (Rosenberg, 1992, p 386). The research organizations are primarily but not exclusively populated by scientists, while the private instrument companies are primarily but not exclusively populated by engineers/technicians. According to Rosenberg, 'the migration of scientific instruments to industry has been matched by a reverse flow of fabrication and design skills that have vastly expanded the capacity of university scientists to conduct research' (1992, p 386). In other words, Rosenberg (1992) understood scientific instrument innovation to be a critical bidirectional link between 'research' and technological 'development'. It was a key point of linkage, coupling, or interaction in the broad process of innovation. This complexity might be reduced to a single one-way line in a transitory model, but there is no doubt that scientific instrument innovation was important to Kline and Rosenberg's theorizing.

Changing lenses

The chain-linked model of innovation is often described as the intermediary step that led from old linear process models to new systems approaches. And thus far I have followed that storyline: push models led to pull models, which led to the chain-linked model and ultimately the systems approach. However, I will now argue that the systems approach and the chain-linked model have overlapping origins: both were responses to the push/pull debate, and both were shaped by an understanding of scientific instrument innovation.

To accept this argument, we must first accept that the 'systems of innovation' approach is not a 'more evolved' version of the linear or chain-linked models. Godin (2017) has dealt with this point. First, he argues that the chain-linked model was essentially another – slightly less linear – process model. As I argued in the previous section, it was a linear model thickly layered with feedback loops. Next, Godin (2017) establishes the difference between linear/chain-linked 'process models' and 'systems models':

> Briefly stated, a process model is one concerned with time, that is, the steps or stages involved in decision making of action leading to innovation (emergence, growth, and development of an innovation). A system model deals with the actors (individuals, organizations, and institutions) responsible for the innovation and studies the way the actors interact. (Godin, 2017, p 5)

This means that process models and system models are different ways of framing innovation. The systems approach does not merely add dimensions (or linkages, interactions, etc.) atop unidirectional models to produce a more complex chain. Instead, it shifts focus away from the work of individual actors (Martin, 2010). This is a qualitatively different perspective on innovation. There are certainly strong arguments for using a systems approach rather than a (linear) process model. But regardless of what these models might give us, the point is that 'process' and 'system' are different lenses for observing the world of innovation. The systems approach is not merely a better process lens; it is a different lens entirely. And in the 1960s, both lenses were being honed through research on scientific instrument innovation.

Focusing on 'systems'

The innovation systems approach is often traced back to the work of Christopher Freeman (1987), Bengt-Åke Lundvall (1988), and Richard Nelson (1993) in the late 1980s and early 1990s. But Lundvall himself goes back even further. He argues that the first articulation of innovation systems theory was Freeman's (1974) analysis of results from *Project SAPPHO* (*Scientific Activity Predictor from Patterns with Heuristic Origins*). Beginning in 1967, SAPPHO (Curnow and Moring, 1968; Rothwell et al, 1974) was the first major undertaking of the newly formed *Science Policy Research Unit* (SPRU) at the University of Sussex. SPRU is perhaps the most famous centre for research on science, technology, and innovation policy (Fagerberg et al, 2012b; Soete, 2019). Christopher Freeman was its founding director and would come to be known as a 'founding father' of innovation studies (for example, Lundvall, 2013a; Martin, 2013; Soete, 2019). This was in no small part due to that first major research project. The SAPPHO results 'attracted much attention, particularly in industry' for both SPRU and Freeman (Fagerberg et al, 2011, p 901). Lundvall argues that Freeman's analysis of SAPPHO was the first recognition of 'the importance of interaction between individuals and departments within firms as well as the important interaction with suppliers, customers, and science institutes' (Lundvall, 2013b, p 41). In other words, SAPPHO provided much of the theoretical foundation for the systems of innovation approach.

Empirically, SAPPHO was a study of science-intensive industrial innovation. The first phase (Curnow and Moring, 1968) was an examination of 58 innovations in chemicals and scientific instruments (see Table 5.1 in Freeman, 1982; and Table 8.1 in Freeman and Soete, 1997). Early on, Freeman was asked to explain this empirical focus. He noted that the work had been influenced by 'the capabilities of the people that were engaged on the study' and by prior studies: work by Enos on petrochemicals and Shimsonhi on scientific instruments (Williams, 1973, p 252). However, he explained that 'there had been no special significance about this [choice of industries] and that they would like in the future to investigate other branches' (Williams, 1973, p 252). Indeed, Freeman initially framed scientific instruments as representing a subset of the broader electronics industry (Freeman, 1973, 1974, 1982). Decades later, Freeman and Soete (1997) would retrospectively explain that scientific instrument innovations were key to the SAPPHO study because they were found at the intersection of fundamental research and new technology development. This seems to be why SAPPHO data were so useful to Freeman in moving beyond linear push and pull logic: much of the data came from observations of pushmi-pullyu devices. However, this is only evident in retrospect. The qualities of scientific instrument innovation were not an explicit part of the SAPPHO design.

The first stated objective of SAPPHO was 'to provide field data for the understanding of the whole process of industrial innovation, and to focus further the search for a better modelling of that process' (Curnow and Moring, 1968, p 82). The research design involved 'the systematic comparison of "pairs" of successful and unsuccessful attempts to innovate' (Freeman, 1974, p 171). Semi-structured interviews were undertaken with the innovating firms and the technological cases were divided into pairs using a qualitative assessment of 'commercial' success and failure (Rothwell et al, 1974, p 259). In other words, the research assumed a market ontology. A scientific instrument could have led to substantial scientific advancement regardless of its movement on the open market. But for the purposes of SAPPHO, scientific instrument failure meant 'never leaving the laboratory' (Curnow and Moring, 1968, p 83). As Freeman explained:

> Since the project was concerned with technical innovation in industry, the criterion of success was a commercial one. A 'failure' is an attempted innovation which failed to establish a worthwhile market and/or make any profit, even if it 'worked' in a technical sense. A 'success' is an innovation which attained significant market penetration and/or made a profit. (Freeman, 1982, p 113)

The resulting analysis identified 27 firm-level factors that differentiated between successful and unsuccessful innovations. Most of the success factors identified by the SAPPHO team were related to marketing practices, and some were related to organizational structure (Freeman, 1974). But according to Freeman, 'the single measure which differentiated most clearly between success and failure was "user-needs understood"' (1974, p 188). He explained that successful innovations were the result of a close 'match' between technology and user needs. He also noted that 'better external communications were associated with success, but the strongest difference emerged with respect to communication with that *specialized* part of the outside scientific community which had knowledge of the work closely related to the innovation' (Freeman, 1974, p 189, emphasis in original). The researchers had asked: 'What was the degree of *coupling* with the outside scientific and technological community in the specialized field involved?' (Freeman, 1974, p 179, emphasis added). And it is in these ideas of *matching* and *coupling* between producers, users, and researchers that we can most clearly see the beginnings of the systems of innovation approach. From these results we begin to get the concept of 'interactive learning', which has been described as the 'theoretical core' of the systems of innovation approach (Lundvall, 2013b, p 32).

Of course, the systems approach is not normally referred to as a theory or model. However, it is certainly of that ilk (Godin, 2017). It has been described as a 'focusing device' – a *kind* of social scientific theory (Lundvall, 1992). It focuses our attention on processes of interactive learning that unfold among actors and within an institutional environment (for example, rules and norms). This framing was more fully developed in Freeman's later book on Japan's national innovation system and in decades of subsequent research on national (for example, Lundvall, 1988; Nelson, 1993), regional (Asheim and Isaksen, 2002; Asheim et al, 2011), and 'sectoral' systems (see Malerba, 2005). In Chapter 6, I will engage with the problem of defining the boundaries for an innovation system and the 'danger of getting "lost in the woods" while searching for the institutional component' (Doloreux and Parto, 2005, p 146). What is important now, in this chapter, is the assertion that 'Freeman's experiences from project SAPPHO provided the ground for the innovation systems perspective' (Lundvall, 2013b, p 41).

Prior to Freeman's analysis, the project seemed poised to take a side in the pushmi-pullyu debate. After all, SAPPHO had been designed in the aftermath of the HINDSIGHT and TRACES research projects – those two large studies that entrenched the push and pull perspectives in the US. It was also conducted in the wake of a high-profile British study (from Manchester Business School) that examined winners of the Queen's Award for Innovation. That study had concluded in support of the demand-pull argument (Langrish et al, 1972; Langrish, 1974). So, when Curnow and Moring (1968) presented the plans for SAPPHO in *Futures*, it was not surprising that they articulated a linear model on the first page. Theirs was a glossy technology push model consisting of three stages: 'technical, industrial and commercial steps, and then the commercial acceptance' (Curnow and Moring, 1968, p 82). In this, SAPPHO appeared like it might become SPRU's counterpoint to the Manchester study, much like HINDSGHT and TRACES were a contrasting set in the US. However, the demand-pull framing suggested by Curnow and Moring (1968) was soon replaced by Freeman's ideas about complexity and systems.

By the time SAPPHO data collection was under way, Freeman had already written some systems language into the first SPRU annual report

(Fagerberg et al, 2011; 2012a; Lundvall, 2013b). And once the first phase of SAPPHO was completed in 1970, Freeman began to clearly articulate his perspective in both the SPRU mission statement (see Fagerberg et al, 2011) and in his analysis of the SAPPHO results (Freeman, 1973). His first written analysis of SAPPHO was presented at a conference of the International Economic Association in 1971 and published in the proceedings two years later (Freeman, 1973). That publication was accompanied by 'minutes' of the conference discussion (Freeman, 1973). There we can see that the lead discussant wasted no time applauding Freeman's approach to the push-pull debate. He said that Freeman's SAPPHO paper 'clearly indicated that single factor explanations were not sufficient to explain success. It was clear that both elements of demand and of research had to be taken into account in explaining the difference between success and failure' (J.J. Paunio, paraphrased in Williams, 1973, p 246). Here, in 1971, Freeman was already moving his peers towards a 'systems' view.

Although the SAPPHO methodology was hotly debated at the IEA conference (Williams, 1973), Freeman would use the results to close debate on the push and pull models (see also Godin, 2017, pp 114–15). His conference paper became the core of his tremendously influential book *The Economics of Industrial Innovation* (Freeman, 1974, 1982; Freeman and Soete, 1997). That book 'for a long time held a virtual monopoly in presenting the "state of the art" of knowledge in the field' (Fagerberg et al, 2012a, p 1136) and had a substantial influence on other major works, notably Nelson and Winter (1982). Starting in the first edition, Freeman asserted that 'innovation is essentially a two-sided or coupling activity' (1974, p 165). He wrote off linear models, saying: 'Whilst there are instances in which one or the other may appear to predominate, the evidence of the innovations considered here points to the conclusion that any satisfactory theory must simultaneously take into account *both* elements' (Freeman, 1974, p 166, emphasis in original). In later editions, Freeman would go further in weighing the SAPPHO evidence against Schmookler (1966) (pull) and 'counter-Schmookler' (push) positions (Freeman, 1982, p 128; Freeman and Soete, 1997, pp 219–20). He would suggest that the push-pull debate was driven by a difference in focus, with one side (push) emphasizing radical innovations and the other (pull) measuring more incremental ones. In his view, the push and pull models simply 'measure something rather different' from each other (Freeman, 1982, p 128).

Interestingly, Freeman's book barely mentions the limits of his own perspective. Those had been raised by the audience at the 1971 conference. And there, Freeman had admitted that the SAPPHO data did not fully account for government–firm relations or the role of public sector organizations as key users of scientific instrument innovations. Nonetheless, 'he admitted that this did have an effect' (Williams, 1973, p 253). He knew that public organizations were important to the successful scientific instrument innovations that had been studied, and that this was not captured in the SAPPHO data or results. Indeed, Freeman's book positioned government only as R&D financier and policy maker. It said 'in capitalist societies most industrial R and D is performed by enterprises and innovations are made by firms' (Freeman, 1974, p 287). There was no space in Freeman's systems perspective for public employees to become user-innovators. And although the SAPPHO results would be cited by von Hippel (1976) in building the case for user innovation, he did not acknowledge the public sector role either.

Neoliberal instruments

Reviews of innovation theory are often punctuated by the insights of von Hippel (1988), Kline and Rosenberg (1986), and Freeman (1974, 1982). Their contributions are central to a canon that is often periodized into three movements: linear, chain-linked, and systems models (for example, Lundvall, 2013b; Martin, 2013, 2016). But in this chapter, we have seen that *material* details have been 'lost in translation'. This was literally true in the case of Joseph Schumpeter and it is figuratively true for these other major figures. When their work is discussed today, the emphasis is on abstracting their theoretical ideas. Ironically, we forget to consider the technologies that shaped their knowledge of technological innovation. We fail to notice that several of the most-cited scholars of innovation shared the same empirical focus (scientific instrument innovation) at around the same time (the late 1960s and early 1970s). I have taken this as an opportunity to write differently about innovation theory.

With scientific instruments as the recurring cast, this chapter has given us a history that is different from the standard canon. The scholars in this story seemed initially unclear about why scientific instrument innovation was so critical to their insights. I have argued that scientific instruments were revelatory because they are essentially pushmi-pullyu devices. Yes, the development of these devices was observed in different ways using wildly different theoretical lenses. But each set of observations challenged the limits of those models; scientific instruments had to be shoehorned into every model. Placing these devices at the centre of this chapter thereby highlights three features of past innovation theory: innovation models themselves are scientific instruments, innovation models share an instrumental neoliberal logic, and innovation research pivots on novel instrumentalities. To conclude this chapter, let me briefly address each of these three points.

Models as instruments

Across several high-profile studies of scientific instrument innovation, we have seen that similar empirical material can be observed using very different epistemic lenses. Observations of scientific instrument innovation were used to support claims about the importance of pushes (Shimshoni, 1966, 1970), pulls (von Hippel, 1976; Utterback, 1971b, 1974), links (Kline, 1985; Kline and Rosenberg, 1986), and coupling/interaction (Freeman, 1974, 1982). This is not to say that any one author's observations were wrong; rather, the mistake is in assuming that theoretical models flow from empirical matters. Models are not merely generalizations or abstractions from material reality. They are also tools for observation and sensemaking. They are instrumental technologies that shape – and are shaped by – our efforts to understand the world around us. The relation between theory and empirical reality is not linear (or even chain-linked).

This is a foreign and potentially invasive concept in innovation studies, where the positivist paradigm still reigns. But it is a rather old idea elsewhere. As I suggested in Chapter 1, STS has a rather more robust understanding of how scientific instrumentalities are woven into science, technology, and innovation. Instruments are highlighted as actors (or actants) in the laboratory studies of Latour and Woolgar (1986) and 'after' (Law and Hassard, 1999). And scientific instruments have been a focal point for many studies in the history (for example, Hughes, 1976; de Solla Price, 1984; Taub, 2011), sociology (for example, Joerges and Shinn, 2001; Shinn, 2005), and philosophy (for example, Marcacci, 2019) of science and technology. These different lines of inquiry share an interest in the epistemology of scientific instruments: a sense that these artefacts defy old Greek ideas about differentiated kinds of knowledge. In other words, scientific instruments are at once *episteme* and *techne* (and, as we will see in the next section, they carry ethical concerns as well).

David Baird's book *Thing Knowledge* is the most expansive of these investigations. He includes models as one form of scientific instrument (Baird, 2004). However, his focus is on physical models and physical instruments. This is because his goal is to outline a materialist epistemology – a sense that 'the material products of science and technology constitute knowledge … in a manner different from theory, and not simply "instrumental to" theory' (Baird, 2004, p 18). And so, Baird establishes a fundamental separation between physical and conceptual models. For Baird, models-as-instruments require maintenance effort, whereas noncorporeal models, like mathematical equations, do not. He argues that conceptual models 'exist in the unchanging, self-sufficient world of ideas' (Baird, 2004, p 35). This is unlike his former colleague Richard Hughes, who was interested in how the diversity of conceptual and physical models 'provide representations of parts of the world, or of the world as we describe it' (Hughes, 1997, p S325). While I appreciate Baird's work, my perspective is closer to that of Hughes (1997): I am approaching models as instrumentalities.

Yes, some theories are primarily tacit. However, this does not make them less powerful actors in the construction of knowledge. Indeed, this chapter suggests that several ideas about innovation – pushes, pulls, chains, and couplings – were powerfully shaping the ways in which prominent innovation scholars understood their empirical observations. These ideas were what my colleague and conference co-adventurer Chris Hartt has called 'noncorporeal actants' (Hartt, 2013; Hartt, 2019). His theory of the noncorporeal actant uses Actor–Network Theory (ANT), historiography, and sensemaking to consider how 'ideas, values, concepts and beliefs are acting upon the decision maker to choose their actions' (Hartt, 2019, p 1). Although his work is focused on noncorporeal actants in past managerial practices, Chris' work helps us position conceptual models alongside other scientific instruments in the production of knowledge. It establishes symmetry between corporeal and noncorporeal actants.

Meanwhile, let's notice that most theoretical models leave physical traces. Some become things (for example, ball-and-stick molecular models) and others are written/illustrated on paper – or mostly on magnetic drives in large server farms. In this way, theories – archived as artefacts, words, and diagrams – are traces of past places and times. Some traces persist and others do not. This is a historicized view of 'extant' theory – the theory available to us today. It recognizes that, over time, knowledge 'which fits with conventional wisdom (not, significantly, with the empirical evidence) is preserved while the rest is truncated' (Jacques, 2006, p 34). The bits and pieces that remain become tools – scientific instrumentalities – for understanding the present. They enable and constrain our present-day thinking.

Instrumental neoliberalism

We get a new angle on the constraints imposed by these innovation models when we look at them as scientific instrumentalities being deployed in the study of scientific instrument innovation. Again, we have the benefit that different models relate to scientific instrument innovation in different ways. We can certainly see how chain-linked and systems models are improvements over 'simplistic' linear thinking. But we can also see how, despite their differences, all the models in this chapter focus on firms and markets. Multiple times, prominent innovation scholars came close to noticing that physical technologies were being developed within universities and public research laboratories. But each time, they turned away. Their models obscured any direct observation of public innovation.

Again, this is not to say that any of these scholars was wrong. Rather, this points to the way in which neoliberalism is 'scripted' (Akrich, 1994) into the shared toolkit of innovation studies. Pfotenhauer and Juhl (2017) have already argued that the push, pull, chain-linked, system, and triple helix[3] models all presume a market ontology. They write that 'the history of innovation models has remained captive to an instrumental dyadic logic that seeks to connect technologies with markets, and that sees the state as both external and subservient to these two poles' (Pfotenhauer and Juhl, 2017, p 87). Godin (2017) agreed, pointing out the irony that it was the success of scientific research during the Second World War that led to the valorization of technological innovation and 'marginalization' of research starting in the 1950s. Later, he expanded on this analysis, arguing that the idea of 'research as a source of progress' now 'competes, for better or worse, with that of the market: that technological innovation is the commercialization of new goods, and the principal agents in the process are firms, not scientists' (Godin, 2020, p 145). But while his historical analysis is more than sound, I think Godin was too soft in suggesting 'competition' between public science and private markets. I prefer the stronger conclusion drawn by Pfotenhauer and Juhl: 'The history of innovation models can be interpreted as one of *systematic exclusion* of the political state and the constituency it serves in favour of a purified, technocratic understanding of what innovation is, what it is for, and who needs to be involved' (Pfotenhauer and Juhl, 2017, p 79, emphasis added).

I share the view that the state is systematically sidelined in innovation studies. This was evident in the way that Freeman (1973, 1974), von Hippel (1976), and Kline and Rosenberg (1986) all asymptotically approached the place of public sector organizations in scientific instrument innovation. But this does not imply any 'dark' intent on their parts. It only confirms that, whether consciously or not, neoliberal politics were embedded in the innovation models that were at play. Models-as-instruments are therefore not only technical and epistemic, but also ethical. However quietly, theoretical models carry values. Although this point might be commonplace for those trained in STS, it is crucial for advancing (critical) innovation studies. I present it here to spur sensemaking about innovation theory – to argue, as Karl Weick (1996) did in organization studies, that these tools are weighing us down.

Instrumentality innovation

In assuming the editorship of *Administrative Sciences Quarterly*, Weick (1996) famously presented the Mann Gulch and South Canyon wildfire disasters as an allegory for the future of organization studies. He described how 27 firefighters died, within sight of safe zones, because they failed to drop the heavy tools that were slowing down their escape, despite direct orders to do so. He then reviewed several slightly less existential threats to organization studies and discussed the 'heavy' scholarly tools impeding progress. His editorial called for a return to 'the lightness associated with "the play of ideas", improvisation, and experimentation', but warned that this would be impeded 'when dropping ideas or keeping them becomes confused with dropping or keeping group ties' (Weick, 1996, p 312). Throughout this book I will argue that his advice and warning apply more acutely to innovation studies.

In this chapter we have seen that innovation models are heavy and potentially problematic tools. We have also seen that they are strongly connected to the identity of innovation studies as a field of scholarship. Indeed, my argument goes further than Weick. I have suggested that innovation models reflect more than shared ideas; they also reflect shared (neoliberal) values. I worry that some might see these values as core to any community of innovation scholarship. They might therefore misconstrue my critique as a personal or community attack. To a certain extent, Godin was right when he said 'the persistence of the market-first perspective speaks more about the values of the scholars promoting it than to its contribution to understanding technological innovation' (2017, p 125). But I take solace in the playfulness of the pushmi-pullyu debate. Yes, this debate was academically fierce and politically charged. However, as we have seen in this chapter, it spurred tremendous scholarly innovation. Studying the 'unusual' field of scientific instruments, from many different perspectives, produced a range of novel ideas about innovation. Models were being retooled left, right, and centre. And so, while the field might now be experiencing 'disciplinary sclerosis' (Martin, 2013, p 179), it also has a history of examining phenomena that do not quite fit and thereby developing new social-scientific instrumentalities. We can draw on that history to address the 'heaviness' of the models explored here and the 'heaviness' of other instrumentalities in the chapters to come.

3

Historiographic Context

'Have you seen the Province's new piece on ocean tech?' a good friend asked me, over beer and pub food one spring afternoon. My mouth was full and so I shook my head to say no. This was the first I had ever heard of ocean technology in Nova Scotia. 'The report is called "Defined by the Sea" and you need to read it', my friend advised. 'They're betting the future on this sector. It's not a real strategy, but there's something there.'

We were at the *Spitfire Arms* pub and this was a brief respite from the first 'two-month intensive' in my PhD studies. There were still two years of coursework to go before comprehensive exams. Nonetheless, people kept asking about my thesis topic. Some things were 'given'. I had studied economic geography in my master's programme. My PhD supervisor did work on innovation systems and had flagged the emerging use of social network analysis methods. I could get in on the ground floor with that method. So, the only outstanding issue seemed to be 'context'.

I grew up in Nova Scotia and had worked here in various local economic development organizations since my teens. The provincial geographical focus was not up for debate (although, it will be in Chapter 6 of this book). The only problem was deciding on a focal sector. I knew quite a bit about the local biotech and software/ information and communications technology (ICT) industries. These both seemed to be 'hot' in innovation research, but I wasn't hot on them personally. Over the preceding months, I had also considered both the wine industry and 'advanced manufacturing' – two sectors of interest for my supervisor. But I was craving policy impact and it sounded like ocean technology might be the next big thing.

I downloaded the new policy document as soon as I was back at a computer. I was expecting something boring – a technocratic assessment of sectoral assets and opportunities. Instead, I was taken aback by passionate rhetoric. *Defined by the Sea* (Government of Nova Scotia, 2012) told a compelling story about a sector that hardly anyone was talking about. It said:

It should come as no great surprise that in Nova Scotia, where the sea has been the defining physical and economic feature for centuries, a strong, dynamic oceans technology sector is well established and growing.

However, the diverse nature of the enterprises and the fact that, on a per capita basis, the province boasts North America's highest concentration of oceans technology companies may raise an eyebrow or two. We suppose that is in part our omission – *a reserved reluctance to tell the story, until now.* (Government of Nova Scotia, 2012, p 2, emphasis added)

I would later learn that this story had been told – with important differences – at least twice in the preceding 50 years. But for now, the 2012 document had achieved its purpose. It was written to convince us that Nova Scotia's new regional competitive advantage would be in ocean technologies. And I was hooked.

History?

Rhetorical history

Defined by the Sea was not intended as a history, but it is. It tells us a story about the past. That makes it a history. But it is not just any kind of history. It is rhetorical history. Companies use rhetorical history to establish valuable symbolic assets and competitive advantages (Suddaby et al, 2010). This involves 'strategic use of the past as a persuasive strategy to manage key stakeholders of the firm' (Suddaby et al, 2010). In this chapter, we will see that rhetorical history also helps assemble a 'cluster', 'industry', or 'sector'. An industrial history can provide a kind of geopolitical competitive advantage (c.f. Porter, 1990; 2003). The narrative serves to attract interest and resources towards the future development of an 'industry'. It establishes the industry or cluster as 'historical fact'. Similarly, Philip Roundy has shown that narratives about entrepreneurship and place (or entrepreneurial ecosystems) can discursively construct regional advantages and disadvantages (Roundy, 2016, 2018; Roundy and Bayer, 2018). He argues that such regional narratives can coalsce over time (Roundy, 2018). But I have shown elsewhere that alternate narratives are always possible and present (MacNeil et al, 2021). Most of us take for granted that there should be one coherent story about a region, industry, innovation system, or entrepreneurial ecosystem (or whatever other container you wish). This seems natural if we accept the most powerful narrative at face value.

As we will see later in this chapter, *Defined by the Sea* provides us with a story about ocean technology in Nova Scotia that is politically motivated and temporally situated. We are meant to accept the version of the past that is presented: without any sense of sources, method, or authorship. After all, the past is merely background context for the *real* work: building the future of this industry. And, make no mistake, industry is the central concern. Public research organizations are cast as supporting characters. What about ocean science instruments? They are just one part of one subcategory of a huge industry – an industry anchored by shipbuilding. In this policy document, the industry and its past are rhetorically constructed.

It's not that the government employees who wrote this report violated the rules of history. Almost everyone 'does history' this way. I did it in my PhD thesis; I just added more facts (and sources). I wrote a straightforward, realist account of the historical context for ocean science instrument innovation in Nova Scotia. 'History' provided the background context I needed to explain and interpret a 'more rigorous' quantitative network analysis. There was no need for historiographic complexities. I just needed readers to accept the past and move on. The context chapter of my thesis was a rhetorical history, just like *Defined by the Sea*. However, this chapter is not.

History as background

Despite 'Schumpeter's plea' that we apply historical analysis to innovation and entrepreneurship (Wadhwani and Jones, 2014), there are hardly any historians in innovation studies (Godin, 2017, 2020). Just like management and organization studies, stories about the past are mostly taken for granted. History is treated as 'background information secondary to the kind of "real" analysis and rigour the social sciences provide' (Wadhwani and Bucheli, 2014, p 7). It is almost always relegated to the short 'context' and 'background' sections that can be found in many studies. There, the past is presented as 'stylized facts' (Kirsch et al, 2014). The emergence of a particular industrial context is an uncited preamble to present-day empirical analysis. For example, influential innovation studies describe the history of nations (for example, Freeman, 2002), regions (for example, Asheim and Coenen, 2005), or industries (for example, Cooke, 2002) without debate over how those histories might come to be known. Yes, time and change are important to innovation theory: temporal processes are at the core of evolutionary economics (Nelson and Winter, 1982) and the concept of path dependence (Liebowitz and Margolis, 1995). But generally, innovation studies do not engage with historiography. Instead, they reproduce what Gaddis (2004, p 92) has called the 'don't ask, we won't tell' approach to historiographic method. They remove all trace of the author and her/his method, and indirectly ask the reader to accept the history, as it is presented. White (1987, p ix) suggests that this writing style gives historical narrative 'an illusory coherence'. It relies on a conventional view that the past is readily available to us.

However, only traces of the past remain: reports, notes, news articles, official and unofficial reports, and so on. We do not have a time machine. The past cannot be retrieved through cleverness of method or comprehensiveness of traces. Traces are simply 'immutable mobiles' (Latour, 1987): ideas that have been inscribed with a certain degree of permanence. They were authored and preserved (by whom? Which ones?). The traces that survive today can interact with other traces, readers, and authors. When we interact with a trace, *translation* (Callon, 1986) takes place. The trace might enrol us into its cause, and we might enrol the trace into our cause. In this way, traces are not objective (or even subjective) evidence of the past. They are nonhuman political actors that can build coalitions around their ideas. When we read a trace, we are given the opportunity to enrol in its network, to act on its behalf, to pass along its story. Traces can connect actors-long-gone with actors-in-the-present. In this way, knowledge of the past is constructed through the relations between actors. Any 'history' can be thought of as an actor-network – or an assemblage. It can be read at face value or disassembled and deconstructed.

This is an amodern understanding of history. It was introduced to me by my friend and colleague Gabrielle Durepos. Her 'ANTi-History' approach fuses ANT with ideas from critical historiography (Durepos and Mills, 2011, 2012). It provides a framework – a set of instrumentalities – for practising history differently. Many critical organization historians now use this approach (for example, Myrick et al, 2013; Peter and Lawrence, 2017; Tureta et al, 2021). It can reveal how knowledge of the past has been constructed and what other ways of knowing the past might be written off, written out, or marginalized.

ANTi-History

Early ANT studies focused on the construction of knowledge in scientific laboratories (Latour and Woolgar, 1986; Latour, 1987) and the social engineering of technologies (Callon, 1986a; Law, 1986). ANT has since been translated for use in organization studies to help retheorize a wide range of topics (Woolgar et al, 2009). But ANT is neither *theory* nor *method* in the traditional sense of those words (Latour, 1999; Law, 1999). It is more accurately described as a research *approach* (Alcadipani and Hassard, 2010).

An actor can be any entity (human or nonhuman, such as a piece of technology) with the capacity to act upon another (Law, 1986). Interactions between these entities form network relations. Callon (1986) calls this process 'translation' and breaks it down into four 'moments'. The interactions begin with *problematization*: a problem is defined (by one or more of the actors) such that the actors seem indispensable to one another. For example, a policy maker might work to convince various organizations that they exist in a mutually beneficial industrial cluster (that is, their fortunes are tied together). The second moment is *interessement*: an actor 'attempts to impose and stabilize the identity of the other actors it defines through its problematization' (Callon, 1986, p 204). In other words, an organization might be convinced that its *identity* is defined through its association within an industrial cluster. Next, objections and negotiations are resolved during the third moment: *enrolment*. Here, some companies may find enrolment to be in their best interests: they benefit in some way from the association. Other companies may choose to resist. In the final moment, *mobilization*, those actors that were successfully enrolled may begin to act on behalf of the network. Company officials may begin to positively promote their home region to suppliers, customers, and partners. Of course, this process is not specific to the formation of industrial clusters.

When many actors begin to act in unison, their network is said to become punctuated (Latour, 1987); in other words, the network becomes an actor. One way in which networks punctuate is by inscribing their intent into a report (Latour, 1987), book (Durepos and Mills, 2012), technology (Akrich, 1994), or other material artefact. This object becomes a nonhuman actor, an immutable mobile (Latour, 1987) that can travel across time and space enrolling other actors into its cause. It also has the ability to appear as a 'black box' (Latour, 1987): concealing the network that led to its creation. Historical accounts are particularly interesting 'black box' inscription points.

ANTi-History (Durepos and Mills, 2011, 2012) is an approach for studying black-boxed organizational histories. It sidesteps the realist/relativist debate in history by providing a 'relationalist' ontological alternative (Durepos, Mills, and Weatherbee, 2012). This means treating historic accounts as knowledges that are embedded within network relations. Mannheim (1936) argued that knowledge must be understood from within the sociohistorical boundaries ('communities') in which it is created. From this perspective, our knowledge of a phenomenon depends on our situatedness. It is a function of our position within a particular network at a particular point in time. It is *relational*. This means that 'two communities can have different knowledge of a phenomenon because of their differing relationships with it' (Durepos and Mills, 2012, p 271).

The vast majority of ANT studies resemble ethnography, where researcher(s) literally follow and observe actors as they relate to one another in real time (for example, Latour, 1987). Meanwhile, ANT approaches are also used with textual data (for example, Callon, 1986a). In particular, ANTi-History research makes extensive use of archival sources (for example, Durepos and Mills, 2011; Hartt et al, 2014; Myrick et al, 2013) to 'follow' actor-network traces. Archival sources were once 'largely ignored' by organizational researchers (Kirsch et al, 2014, p 235). However, Kirsch et al called for researchers to engage with archival material 'in order to create new analytical narratives of industry formation' (2014, p 235).

I began my own archival work from the idea that *Defined by the Sea* (Government of Nova Scotia, 2012) had 'revealed' the neglected history of a vibrant ocean technologies cluster. This inspired me to search for older historical traces. I began by searching for 'ocean technology/ies' at the Nova Scotia Public Archives. I later added material from the Dalhousie University and the BIO libraries and archives. I examined a total of 70 documents dating from 1944 to 1995, including official and unofficial government/ agency reports, books, newspaper clippings, and magazine articles. Over approximately one month of research, I produced 58 pages of notes and recorded 60 pages of annotated images. This work was not necessarily linear and chronological: an interesting point found in the 1980s would cause me to re-examine traces from the 1970s.

Among all the data I collected, three documents stand out for their attempts provide a comprehensive (historical) account of an 'oceans cluster' in Nova Scotia. Other related archival traces helped me to understand how each of these three accounts was situated within different actor-networks (or, different temporal and relational contexts). Let me begin with a closer examination of the most recent account – the one that my friend told me about that day over beer at the pub. Then, I will work 'backwards' through time, sharing accounts from 1980 and 1960. As we will see, these three histories disagree on the 'context' around ocean science and technology in Nova Scotia. Once I have explored each of them in their own time, I will turn to what we might learn about 'practising context' based on the tensions between these accounts.

Three historical accounts
Defined by the Sea

In the summer of 2012, the Government of Nova Scotia published *Defined by the Sea: Nova Scotia's Oceans Technology Sector Present and Future*. This document was posted on the website of the Department of Economic and Rural Development as part of the governing New Democratic Party's *jobsHere* initiative. This meta-initiative was a widely promoted job creation programme launched in the autumn of 2010, which served as a cornerstone of the New Democratic Party's unsuccessful re-election campaign in 2013. The *Defined by the Sea* document positioned 'ocean technologies' as a priority sector within the *jobsHere* strategy.

The short 24-page report begins with ten 'at-a-glance' bullet-points about the sector. The first of these is a claim that the sector includes: 'Over 200 companies. More than 60 innovators of new high-tech products and services' (Government of Nova Scotia, 2012, p 1). The bullets also tell us that these companies have combined corporate revenues of $500 million, perform one third of all private research and development in the province, and pay nine times more taxes than the average Nova Scotian firm. Four major public research organizations are highlighted: the BIO, the National Research Council's Institute for Marine Biosciences, the Defence Department's Defence R&D Canada, and Dalhousie University. The mention of these institutions follows the claim that 'Nova Scotia is home to 450 PhDs in oceans-related disciplines. Highest concentration in the world' (Government of Nova Scotia, 2012, p 1). The final bullet in the summary addresses opportunities for sector growth: 'Estimated annual global market value for ocean-related goods and services: US$3 trillion. Doubled in last six years' (Government of Nova Scotia, 2012, p 1).

While *Defined by the Sea* does not describe the 'evolution' of this industry in historical terms, I have said that it is a history. Indeed, the document reads like a cross between an industrial cluster analysis and a glossy promotional booklet (complete with professional photographs of key private sector ocean activities). The text includes a general introduction to 'the sector', descriptions of successful companies, an assessment of market opportunities, a summary of public research capacity, and a description of the 'enabling environment' (such as training institutions, industry associations, and government funding programmes). It also includes an extensive collection of impressive 'facts' (without mention of their origin or authorship). The most tenuous of these relate to the size of the industry. The narrative assembles a wide range of companies into this 'sector' largely due to the broad definition used: 'The oceans technology sector comprises "knowledge-based companies that invent, develop and produce high tech products *for specific use in or on the ocean*; or provide knowledge-intensive, technology-based services, *unique to the ocean*"' (Government of Nova Scotia, 2012, p 2, emphasis added). Including all ocean-related knowledge-based companies means that the reader will encounter many unrelated firms. For example, the text discusses a recreational boat builder, a nutritional supplement company, and a naval defence sonar manufacturer. To accommodate this diversity, the sector is grouped into six 'key areas of concentration': 'acoustics, sensors, and instrumentation; marine geomatics; marine biotechnology; marine unmanned surface and underwater vehicles; marine data, information, and communications systems; and naval architecture' (Government of Nova Scotia, 2012, p 5). This definition provides for the claim that Nova Scotia has 'North America's highest concentration of oceans technology companies' (Government of Nova Scotia, 2012, p 2).

Defined by the Sea was written at a time marked by both government austerity and targeted economic stimulus. It was published one year after Irving Shipbuilding was selected to build Canada's next-generation naval vessels at its Halifax dockyard. To this end, the Government of Nova Scotia had led a $1.4 million, widely criticized lobbying and public relations campaign entitled 'ShipsStartHere' (McLeod, 2011). *Defined by the Sea* reads like an extension of that public relations effort. In short, it tells the reader that an already sizeable industrial sector – one connected with shipbuilding – is poised for significant future growth. This is encapsulated in a quote from Premier Darrell Dexter on the final page of the document:

> Take the collective strength of oceans-related research capacity in the province; combine it with the proven entrepreneurial vision of Nova Scotia's oceans technology leaders and companies; add committed government support and promotion and the opportunities for economic growth are limitless; the solutions to some of the most vexing problems of our time are within reach. (Government of Nova Scotia, 2012, p 21)

Interestingly, oceans-related opportunities had also been within reach 30 years earlier.

'One of the three biggest'

Canadian Geographic ran a 12-page feature story about Nova Scotia's 'marine science cluster' in its October/November 1980 issue (Watkins, 1980). The headline reads 'Halifax-Dartmouth area: one of the three biggest marine science centres in Western Hemisphere' (Watkins, 1980, p 12). The author goes on suggest that this 'probably is the third largest (if not the second largest) concentration of marine research and development facilities and personnel in North America' (Watkins, 1980, p 13). His claim is highlighted in a pull-quote: 'Public and private enterprises oriented to the sea have assembled the largest concentration of marine scientific and technical personnel to be found in Canada, outnumbered in the Americas only by the Boston-Woods Hole area in Massachusetts and perhaps the Scripps Institution in California' (Watkins, 1980, p 12).

This story places a great deal of emphasis on the 'evolution' of the cluster. On the first page, the author invokes a sense of loss over the region's shipbuilding/sailing history. He claims that 'marine science' is restoring Canada's 'maritime reputation' (Watkins, 1980, p 12). The article's purpose is framed in this way: 'This is the story of that renaissance, a look at some of the scientific and developmental involvement behind Canada's rise to international oceanographic prominence. It is an accomplishment *far better known abroad than it is at home*' (Watkins, 1980, pp 12–13, emphasis added). The author situates the origins of his 'renaissance' in the establishment of a naval defence research unit during the Second World War. His focus, however, is on the work of the federal government's BIO. He explains that: 'The Bedford Institute of

Oceanography is at the core of a cluster of marine-oriented establishments which includes the Nova Scotia Research Foundation, Dalhousie,[1] and the newly-named Technical U. of N.S.' (Watkins, 1980, p 12).

The article is nearly exclusively focused on these public research organizations. For example, it speaks of the BIO-led first-ever circumnavigation of the Americas, aboard the scientific vessel *CSS Hudson*[2] (in 1970). There is also a lengthy section about ongoing work to assess the possible ecological impacts of tidal power dams. The text is accompanied by photographs of scientists and scientific activities, including a crew lowering a rock-core sampling drill into the ocean. A few general photos of Halifax harbour are included, with captions that point to the location of key public research buildings.

However, private sector activities do not go completely unnoticed. There are mentions of spin-off technologies arising out of the research laboratories and their activities. For example, the article says:

Most of BIO's deep-ocean research is conducted with the aid of instruments devised and engineered in the Halifax area. These include the Batfish, a remarkable device which is towed behind a ship and which dives and climbs to varying depths, automatically gathering information on such things as temperature, salinity, conductivity, light, and chlorophyll fluorescence. It can also catch samples of small animal plankton. Developed at BIO, it is made in Smiths Falls, Ont. (Watkins, 1980, p 22)

Notice the focus on how this device was developed and used by the scientists. They are being supported by an unnamed organization (Guildline Instruments) in Ontario.

There is also a strong focus on offshore (Arctic) oil and gas opportunities at the end of the article. This is presented as a potential stimulus for the cluster's future growth. The emphasis on petroleum development is partly explained by the author's bio: 'as a result of his interest in shipping and marine affairs he was aboard the tanker Manhattan during her pioneer voyage through the Northwest Passage in 1969' (Watkins, 1980, p 12). But it is also partly explained by BIO's research agenda, which had long been devoted to Arctic exploration (for the purpose of Canadian Arctic sovereignty) and had become particularly focused on Arctic petroleum development throughout the 1970s (Nichols, 2002).[3] Similarly, the Nova Scotia Research Foundation (NSRF: a crown corporation devoted to applied research for economic development) had become focused on offshore petroleum resources. When this article was published, the NSRF had been led for over a decade by Dr J. Ewart Blanchard, a marine geophysicist, who was recruited from the original faculty of Dalhousie's Oceanography Institute.

Notably absent from the author's description of the cluster is naval defence. Its role is relegated to activities during the Second World War. The only sense of Cold War tensions in this article arises in a discussion of BIO's work to detect nuclear waste disposal in Canadian waters. The absence of naval research became clearer when public archivist Rosemary Barbour helped me locate 'Knots, volts and decibels: an informal history of the Naval Research Establishment, 1940–1967' (Longard, 1993). Written in 1977, the foreword to this booklet explains that the Department of Defense deemed it too sensitive to publish until the 1990s. This suggests that the Defense Department R&D division may have remained an unmentioned actor in the *Canadian Geographic* article because its work was highly secretive at that time (for more on this, see Chapter 4).

'Internationally important'

Canadian Geographic's discovery of a marine science cluster in Nova Scotia was pre-dated by a similar discovery in the local newspaper a decade earlier. On 6 August 1960, the *Chronicle Herald* proudly proclaimed: 'Halifax is becoming an internationally important base for one of Canada's biggest tasks – the oceanographic study of her virtually unexplored northern waters' (Trenbirth, 1960). This article is presented as an origin story: it speaks of how Halifax is 'becoming' an important region for a field of science that is 'in its infancy' (p 6). It describes in detail the vessels, personnel, and technologies that left Halifax harbour that summer to conduct oceanographic research. It notes the use of 'radar screens' and 'echo sounders' and 'a trail of moored detectors. Left bobbing on the surface they will self-record information while the ship continues its trip' (Trenbirth, 1960). Individual scientists and ship's crew are applauded for their skill. The key organizational actors in this story are arms of the Government of Canada: the Canadian Committee on Oceanography and its Atlantic Oceanographic Group.

Seven months earlier, the federal Department of Mines and Technical Surveys had announced $3 million (approximately $30 million in today's Canadian dollars) to build BIO, on the advice of the Canadian Committee on Oceanography (BIO, 1962–92; 'Canadian Institute of Oceanography', 1959). The facility would be home to various federal government departments and agencies engaged in fisheries and oceans research. The previous year, $90,000 in federal funding (equivalent to approximately $950,000 Canadian dollars in 2023) had been announced to establish an oceanography institute at Dalhousie University (Hayes, 1959). Dalhousie had been lobbying heavily since 1949 when the University of British Columbia secured federal funding for its own west coast oceanography institute (Mills, 1994). These two major funding announcements made ocean research particularly noteworthy in Nova Scotia during the summer of 1960. The *Chronicle Herald* article directly addresses the future impact of this funding with an anecdote about research on tidal currents in the Bay of Fundy: 'their instruments were unable to record the rapid tide flow, estimated at eight knots. But when the Oceanographic Institute at Bedford Basin gets down to business in 1962, they hope to manufacture their own instruments instead of importing them from Europe and America' (Trenbirth, 1960, p 6). Here we see a desire for Canadian economic and technoscientific sovereignty; this was the same decade that saw the repatriation of the Canadian Constitution from the British government. And through anecdotes like this, the *Chronicle Herald* article also articulates the anticipation that some Nova Scotians felt towards these new ocean research institutes and a possible research cluster. As BIO's 1963 annual report would later state: 'It seems doubtful if even the new Dartmouth brewery will be more warmly welcomed' (BIO, 1962–92, vol 1963, p 1).

As in the *Canadian Geographic* publication, defence research remains notably absent from this article. There is a brief reference to military research interests, but these are quickly brushed aside. A naval vessel, the HMS *Sackville*, is an important actor in the story, but she is an 'auxiliary' vessel and therefore deployed for science rather than defence. Cold War tensions are explicit in the article, but they are described in terms of scientific supremacy. This begins in the second sentence: 'Canada and the United States have merged forces. The Russians also have been in this port' (Trenbirth, 1960, p 6). The author names two Soviet ships that called on Halifax harbour earlier in the year. Then, near the end of the article, Russia is described as having a competitive advantage over Canada: 'Canada, with its long coastlines, has a lot of leeway to make up. Russia years ago seized on to the importance of oceanography' (Trenbirth, 1960, p 6). This article positions ocean science as an international competition. Meanwhile, several annual reports from the BIO (1962–92) present similar visits by Russian research vessels as international collaboration.

In addition to the absence of naval research, a notable actor from the *Canadian Geographic* history is also missing from the 1960 account. The NSRF was already well established (having been founded in 1946) when Trenbirth's (1960) newspaper story went to press. However, its annual reports demonstrate that it was first focused on agriculture, mining, and fisheries The NSRF's oceans research would not begin in earnest until later in the 1960s (NSRF, 1946–95), following the lead of these other institutes. The provincial government therefore does not appear to be a significant actor in this early network. Instead, the federal government, and its funding of scientific research, is the central actor. Note that this article was published at a time when the Canadian and American governments were both making significant investments in science for the purpose of stimulating industry development (Doern, Castle, and Phillips, 2016).

Context?

Interessement

Ten years ago, I went to the Nova Scotia Public Archives hoping for enough material to write a simple context chapter. That work was complicated by these three accounts of the past – each constructed in (and now abstracted from) a different temporal context. At each of these three points in time (1960, 1980, and 2012), the reader is informed that historical processes have *recently* pulled together a cohesive oceans cluster. In *Defined by the Sea*, the reader is previously unaware of the cluster because the government has 'omitted' it from history, due to 'a reserved reluctance to tell the story, until now' (Government of Nova Scotia, 2012, p 2). In the *Canadian Geographic* article, the reader is unaware because the cluster is 'far better known abroad than it is at home' (Watkins, 1980, pp 12–13). And in the *Chronicle Herald* article, the reader is unaware because she/he is assumed to have missed major funding announcements over the preceding year (Trenbirth, 1960).

These repeated attempts to establish a coherent cluster identity are made possible through the failure of similar attempts in previous decades. These appear to be failures in translation (Callon, 1986). The first moment of translation, problematization (Callon, 1986), was similar in all three of the accounts I studied. Each author argued that the cluster's existence was going unnoticed. They all rhetorically positioned the cluster as one of the biggest and best in the world. Furthermore, they all argued that the cluster was on the cusp of tremendous growth. It is presented as a point of pride for those involved and for Nova Scotians at large. Many public and private ocean science organizations are drawn into these problematizations.

This provides for the second moment of translation: *interessement* (Callon, 1986b). Here, the authors each impose a collective identity on the characters in their stories. While each story uses similar words for the 'ocean science' (and technology) cluster, they each define the 'contents' of that cluster differently. The sectoral/industrial boundaries are produced through the rhetorical devices used by each author to include/exclude the actors that may align with their cause (for more on this, see Chapter 6). Similarly, the geographical boundaries for the cluster are an interest-driven choice. The *Chronicle Herald* article (Trenbirth, 1960) uses 'Halifax', the *Canadian Geographic* article (Watkins, 1980) uses 'Halifax-Dartmouth area' (the harbour is the area of focus), and the government report (Government of Nova Scotia, 2012) uses the entire Province of Nova Scotia. These geographical labels not only help to explain which actors might be inside or outside the network/cluster, but are also a form of geopolitical identity work (for more on this, see Chapter 6). The authors attempt to make this cluster a part of the provincial, or at least capital city, identity. This form of rhetorical positioning then becomes a symbolic asset: one that might be used to generate regional competitive advantage.

However, the translation process does not seem to have progressed beyond *interessement* for very many actors over the past 50 years. Perhaps the various 'ocean cluster' identities simply could not be negotiated among such a diverse set of actors? Actors need to become enrolled in the network for each 'cluster' to exist 'beyond the page'. On another level, readers need to become enrolled in a history in order for it to be accepted and then passed along. However, in these three examples we observe a singular cluster that is (re)discovered, with notable differences, time and time again. These histories seem unable to enrol and mobilize actors. The 'cluster' they purport to discover is therefore unable to remain black boxed outside pages of these histories. It 'disappears' into bits and pieces that await (re)assembly at some future point. Along the way, some bits and pieces of story are lost, and others are set aside.

Elision

Astute readers will have noticed parallels between this chapter and the last. In Chapter 2, I noted the disappearance of scientific instruments from their privileged position in innovation theory. Now, in this chapter, we have seen the disappearance of ocean science instruments from their privileged position in Nova Scotia's history. The earliest account of the sectoral history (1960) positions scientists, scientific instruments, science organizations, and geopolitics as key actors. But in the latest account (2012), scientific instruments are marginal; the main actors are private companies (especially shipbuilding), and science is described as providing supportive human capital. No doubt, the lessons from Chapter 2 apply here: *Defined by the Sea* (2012) was written with a mental model connected to the ideas of industrial clusters and regional innovation (eco)systems. As we have seen, those models centre on businesses and markets. They position science in the background, base, or foundation of an industry's evolution. Kirsch et al (2014) explain that such models can shape the stories we tell about industry emergence. Missing pieces of story can be 'the result of a certain retrospective myopia that comes from imposing an extant social scientific explanation on evidence from an industry's past' (2014, p 228).

But let's not overstate the case. It is not that ocean science and scientific instrument innovation have been forgotten or lost to time in Nova Scotia. Scientific actors have not disappeared; they have been repositioned. While they were central to the accounts of 1960 and 1980, they were marginalized in favour of more important actors in 2012. And so, the full scope of their role has been elided, but their presence has not. These changes in the narrative are related to changes in the dominant social scientific models. But the historiographic definition of an industry and its evolution is a practical problem before it is a scholarly one (Kirsch et al, 2014). The cognitive view of industries tells us that individuals have their own sense of what industry they are participating in, what companies and products it contains, how these are defined/categorized, and which pieces of the industrial puzzle are the most valuable (Porac et al, 1995; 2011; Kirsch et al, 2014; Khaire, 2014). Theory and mental models are one thing, but the broader and more important questions are related to cultural and historical context.

We cannot understand stories of industrial evolution apart from the context in which they are written (Kirsch et al, 2014). These stories claim to provide 'context', but they also *have* context. Therefore, strong analysis of industrial history must engage with hermeneutic philosophy (Kirsch et al, 2014). In short, the hermeneutic insight is that text and context are circularly linked (Prasad, 2002). Text is always situated in context; context is always constructed from text. Kirsch et al (2014) explain that actors'

> perceptions of industry formation are shaped by expectations established by their own position in historical time, as they look back from the present into the past. This retrospective view *reorders* what we observe about industries and their emergence in light of both subsequent developments and the conceptual biases we hold, while emphasizing what we understand to be of *post hoc* importance while marginalizing other developments in the industry emergence process. (Kirsch et al, 2014, p 222, emphasis in original)

Traces of the industrial past are elided and 'facts' are stylized based on our sense of what is important now rather than what was important then (Kirsch et al, 2014). In other words, we are situated in a cultural context whenever we write or read about the past. This is historicism (see Kirsch et al, 2014). The alternative – *presentism* – is the bias we apply when we read and write about events of the past with current cultural lenses. It is an instrumental bias.

To avoid this, we must do more than question our own interpretive prejudices. Critical hermeneutics also suggests that we should deconstruct ideology found within text and context (Prasad, 2002). On that note, it would be easy to explain away differences in the 1960, 1980, and 2012 stories earlier in this chapter with 'the rise of neoliberalism'. But good historical analysis also requires a more situated explanation. Here, ANTi-History helps us avoid getting lost in generalities and the circular logic of interpretivism. It provides an alternative analytical metaphor (or an alternative instrumentality): advising us to rhizomatically 'follow the actors' (Latour, 2005). This allows us to deconstruct our own positions, and the positions of the actors/traces we engage with in our research.

For example, let's deconstruct the 'neoliberal' explanation by examining how public and private organizations were enrolled into these different 'oceans cluster' histories. The most recent account (2012) tells us that naval defence is, and *has always been*, a critical part of the cluster. However, this role is muted or absent in the earlier accounts (1960 and 1980, respectively). In the oldest account, the federal government (in its role as research patron) and its *Canadian Committee on Oceanography* are the central players in a relatively small network of actors (research vessels, technologies, and skilled experts). By 1980, the BIO has become central to a cluster of predominantly public sector organizations. Then in 2012, over a dozen private sector companies are named as main characters in the story and several public research organizations are named as a supporting cast. In short, the earliest account expressly focuses on the role of federal government science organizations and funding in establishing the cluster, and this goes unmentioned in the most recent of accounts. In both 1980 and 2012, private markets are providing the cluster with its capital. This contrast, between a predominantly public and predominantly private sector characterization, is likely explained by the rise of a broader neoliberal economic discourse in Canada over this period (Carroll and Shaw, 2001). But we can follow the actors to a more specific and situated explanation. Indeed, *Defined by the Sea* was written at the height of neoliberal attacks on science in Canada. Rather than taking its narrative at face value, we must try to understand *Defined by the Sea* by what it omits from its own historical context.

Cutting science

Chris Turner argues that throughout the early 2010s, the Government of Canada exhibited 'mounting disdain for the work of its scientists' (2013, p 17) and enacted 'vicious cuts' (2013, p 26) to public research organizations, particularly those within the Department of Fisheries and Oceans. He suggests that this 'war on science' was punctuated by the 2012 federal budget legislation, Bill C-38:

> No scientist working on a federally funded project in the spring of 2012 could have been wholly complacent about their job security, especially if their field was in the environmental sciences. Bill C-38 had unleashed a broad frontal assault on the Canadian environmental science community. Tabled in the House of Commons six weeks earlier, the bill had triggered wave after wave of closures and 'affected letters' (notices of potential or impending layoff) at research institutes, monitoring stations, and government labs across the country. (Turner, 2013, p 8)

Science journalist Hannah Hoag describes these cuts as a policy shift away from basic science and towards applied partnership with industry (Hoag,

2011, 2012, 2013). This is further to a global trend in science policy (Sá and Litwin, 2011; Archibugi and Filippetti, 2018). Bailey et al (2016) have also suggested that leadership in the Canadian government was attempting to 'devolve' ocean science activities to universities and the private sector. They explain that these cuts served to 'eviscerate Canada's federal aquatic science programs – staff reductions, closures of laboratories, closures of marine science libraries, and cessation of key research programs' (Bailey et al, 2016, p 1). This, too, is further to a global trend in science policy – a shift of resources towards university research labs and away from other types of PROs (Salazar and Holbrook, 2007; Archibugi and Filippetti, 2018).

Turner argues that this shift in policy was grounded in the belief that 'the purpose of research – of science generally – is to create economic opportunities for industry, and the purpose of government is to assist in that process in whatever way that it can' (Turner, 2013, p 112). He describes movement away from 'the open spirit of scientific inquiry' (2013, p 132) towards the view that 'government's job is to deliver innovations like theatre tickets to the front desk of a posh hotel' (2013, p 112). In other words, those in power came to believe that PROs exist to serve the market. This belief has been linked to New Public Management (NPM) practices – a particular set of neoliberal strategies that were popular across Organisation for Economic Co-operation and Development (OECD) countries for decades.

NPM is a label applied to a 'set of broadly similar administrative doctrines which dominated the administrative reform agenda in many of the OECD group of countries from the late 1970s' (Hood, 1991, pp 3–4). The NPM agenda has been studied and critiqued as a set of organizational innovations in the public sector (Schubert, 2009; Hansen, 2011; Lorenz, 2012). NPM has been directly linked to neoliberalism because NPM reforms were intended to make public organizations more business-like (Atkinson-Grosjean, 2006; Lorenz, 2012). While the effectiveness of NPM is debatable (Hood, 1991; Schubert, 2009; Lorenz, 2012), it is accepted that NPM reforms dramatically changed the management and organization of public science in Canada beginning in the 1980s (Atkinson-Grosjean, 2006) and 1990s (Smith, 2004). In these previous waves of reform, public science was reorganized and increasingly aligned to private interests (Atkinson-Grosjean, 2002, 2006). The recent wave of reforms has resulted in substantive cuts to PROs across Canada and particularly to those that were engaged in ocean science (Turner, 2013). Daniele Archibugi and Andrea Filippetti discussed these 'neoliberal forces' (2018, p 98) and argued that the decline in public science globally will have 'long term adverse consequences' (2018, p 12) for development.

It is surprising that these consequences are not front and centre in *Defined by the Sea*. Indeed, that document was written during a 'critical juncture' (Mills, 2010) where science policy and industrial policy were moving in opposite directions. The worst federal cuts to ocean and environmental science were happening while all levels of government were investing heavily in 'ocean technologies' as a priority sector for economic development. Hot on the heels of *Defined by the Sea*, ocean technology became a priority development sector for the City of Halifax's economic development agency (Greater Halifax Partnership, 2012). It had already been a priority for the federal government's multi-provincial Atlantic Canada Opportunities Agency (ACOA) (Atlantic Coastal Zone Information Steering Committee, 2006). In their rhetoric, these various governmental organizations appeared blissfully unaware of the implications the war on science might have for their industrial aspirations. Ocean science had come to be understood as a support system for ocean industry. Ocean science instruments were interesting due to their technical sophistication and market potential. But they were not seen to be as important as other activities, particularly naval shipbuilding. As we have seen, this was quite different in 1960 and 1980.

Practising context

In this chapter, I have tried to show that how we read and write about context matters. This would be a drastically different book if I had taken *Defined by the Sea* at face value or if I had begun my work in 1960 or 1980 (or at any other juncture, with any other narrative). I could have written a much simpler chapter by choosing one storyline. But I have been interested in how knowledge of this 'context' has been constructed and what that might leave out. I agree with Kirsch et al (2014) that we need strong historical reasoning in the study of industrial evolution. As they say, the tools of historiography 'can highlight aspects of industry emergence and evolution that are systematically left out or elided both by the passage of time and by our own social scientific models of industry evolution' (Kirsch et al, 2014, p 218). Choices must be made 'as both contemporary actors and subsequently the social scientists studying them come to focus on certain knowledge as constituting the industry' (Kirsch et al, 2014, p 219). We strip away the complexity of these choices when we take context for granted (McLaren and Durepos, 2019).

Like much of the research in organization studies, economics, and elsewhere, innovation research typically presents context as brief background material. It is mostly framed as a 'fixed container, broad environment, or macrolevel feature' of the phenomenon we choose to study (McLaren and Durepos, 2019). Trish McLaren and Gabrielle Durepos have written against this naïve conceptualization of context. They problematize the tendency to 'either ignore context or treat context in a way that assigns it fixity and immutability' (McLaren and Durepos, 2019, p 78). When we do that, context becomes 'just another variable which is isolatable and thus stripped of its complexity and fluidity' (2019, p 78). They argue that 'comprehensive explanations of context are rare. Also rare are the ontological, epistemological, and methodological considerations that should accompany a rigorous consideration of context' (2019, p 76). There is little space (or appetite) in our scholarly texts for rigorous examinations of context.

But centuries of hermeneutic philosophy have argued that text and context are inextricably linked. Indeed, text and context are only separate in our writing. We choose to write one phenomenon into the foreground and others into the background. These choices are situated. The choices have their own context which must also be considered. Like me, you might fear getting lost in this hermeneutic circularity. And yes, a purely interpretivist approach could lead us to wreck our ships on the 'reef of solipsism' (Sartre and Richmond, 1956) – a place where all text and context are relative. This is likely no better than the purely realist view that the 'facts' of history *must be* clear and indisputable.

ANTi-History has given us a path out of this quagmire. I do not say this only because Durepos is a friend. I say it because an actor-network approach (or actant-rhizome ontology, if you prefer that metaphor) treats knowledge as relational. Rather than attempting to determine the 'truth' of different traces and accounts, discounting 'inaccuracies', and writing a 'better' version, we can take note of the different actors and relations present. In this way, we can understand each of the three histories in this chapter as valid *relational* and *situated* accounts produced by different actor-networks. Here, context is ontologically 'multiple' (Mol, 2002).

This chapter is yet another relational and situated account. I have invited you to follow my (de)construction of Nova Scotia's ocean science and technology past. We began on the day I first learned of *Defined by the Sea*. We entered the archives to locate and understand prior accounts. We then returned to consider the silenced context of the 'war on science'. This was not meant as a self-aggrandizing adventure. Rather, I have tried to 'practice context' (McLaren and Durepos, 2019) and to demonstrate a potentially useful historiographic instrumentality. I submit that this is an alternative to taking context for granted.

4

Narrative Politics

It was almost one year after I had 'discovered' ocean science and technology in Nova Scotia and I was now sitting in a packed conference hall at the Halifax Marriott Harbourfront Hotel, waiting to hear from a panel of prominent local scientists and entrepreneurs. The 'Oceans Panel' was positioned as a key element of *The Premier's Innovation Summit* – a 2013 conference about the future of Nova Scotia's economy. Premier Darrell Dexter was several months away from calling the next election. So, he was working hard to glean political capital from his government's job-creation strategy – particularly from the upgrades that had just begun at Halifax Shipyards. Irving Shipbuilding had won the $25 billion contract to build Canada's newest naval vessels in Halifax. Construction of the vessels was set to begin the following year.

Amid all this enthusiasm for shipbuilding, I expected one of the conference panellists to recentre the discussion on scientific instruments. Dr Marlon Lewis was Chair of Oceanography at Dalhousie University and had been Chief Executive Officer (CEO) of the Halifax-based ocean technology company, Satlantic. Satlantic was one of two important scientific instrument companies to have recently spun out from Dalhousie's Oceanography Department. It manufactured and sold a range of sensors, including a device for measuring photosynthetically active radiation (as in the light used by phytoplankton for photosynthesis) and software for processing the sensor data. Two years earlier, Dr Lewis had sold Satlantic to the US company Sea-Bird Electronics – a merger/acquisition that created the multinational ocean science instrument company Sea-Bird Scientific.

On the panel that day, Dr Lewis did share some of his experiences in the recent history of ocean science technologies, but I was especially taken with his glossy sectoral origin story. From his perspective, the evolution of the ocean technology industry in Nova Scotia can be traced directly back to the Cold War era search for Soviet submarines.

This struck me as an exciting story to tell, like the stories of Cold War-era innovation in the machine tools, commercial aircraft, and information technology sectors in the US (Mowery, 2009). I do love a good Cold War thriller. And yet, as I have argued in the past two chapters, any one coherent storyline is bound to leave things out (just as this chapter surely will).

Narrative devices

In Chapter 3, we saw early histories of ocean science and technology innovation in Nova Scotia that foregrounded science. However, in 2012, science was being characterized as a support system. It was being celebrated primarily for the support it provides to private commercialization, not for any public good it might create. This is not unusual. Indeed, the apolitical treatment of science and technology (Fagerberg et al, 2012b; Martin et al, 2012) – including the apolitical treatment of the state (Pfotenhauer and Juhl, 2017) – has been flagged as an issue in innovation studies. Sebastian Pfotenhauer and Joakim Juhl have argued that 'under the neoliberal paradigm, every public good is captive to the logic of the market, every action is evaluated in terms of return on investment, and state intervention is only justified to rectify market "failures"' (2017, p 88). What we need, they say, is research that centres public organizations and their enactment of innovation. In this chapter, I do this narratively. Using a CMS approach to narrative analysis (Boje, 2001; Czarniawska, 2004; Vaara et al, 2016), I present three short stories that are each centred on a public organization we encountered briefly in Chapter 3. What we get are three different enactments of ocean science and technology in Nova Scotia. Taken together, these three stories resist some of the normal ways we narrate innovation. But before we get to those short stories – my *petits récits* (Lyotard, 1984) – let me briefly explore how the neoliberal metanarrative shapes stories of innovation.

Narrative neoliberalization

Grand narratives are hegemonic frameworks for understanding our world. François Lyotard developed the notion of metanarrative in *The Postmodern Condition* (1984), a piece on the philosophy of science and technology that was commissioned by a group of universities in Quebec. There, he argued that postmodernism is defined by a scepticism towards metanarrative. It is a scepticism towards stories that are reductionist and universal. Such stories are glossy simplifications that leave little room for alternative claims. For example, Lyotard examined the 'Enlightenment' metanarrative of Western science and its uncomplicated and universal claims about knowledge and truth. The Enlightenment metanarrative is present in innovation studies. Several other metanarratives are also worth mentioning in a broader critical innovation studies agenda. I will make some suggestions in Chapter 8. But for now, let's focus our attention on the metanarrative effects of neoliberalism.

At the outset of this book, I argued that neoliberalism is a grand narrative of concern within innovation studies. It is not one coherent idea, but rather a set of variegated political values that have persistent power (Peck, 2010a; Peck and Theodore, 2019). So, we can think about neoliberalism as a 'messy grand narrative' (Phelan, 2007, p 328) or as a variegated process – that is, 'neoliberalization' (Peck and Tickell, 2002; Brenner et al, 2010). This positions us to deconstruct the *narrative neoliberalization* of stories about innovation. Here, I am not accusing any individual authors of being deliberately and secretly political when they write about innovation. I am not even suggesting that neoliberalism is a coherent political belief. Rather, I am saying that narrative neoliberalization is a default – or 'zombie' (Peck, 2010b, p 104) – approach to writing about innovation. It is simply the norm for stories about innovation to feature businesses as the main characters and for those stories to resolve in market success. In other words, neoliberalism is metanarrative.

Restorying analysis

What other stories and voices are set aside in the process of narrative neoliberalization? The postmodern response is to 'restory' the grand narrative (Boje, 2001, p 10). This involves taking multiple 'local stories' (2001, p 35) and assembling them in ways that resist or 'shatter' the grand narrative. The bits and pieces of local story are what David Boje (2001, p 7) calls 'antenarrative'. He uses 'ante' as a double entendre: an antenarrative is both a precursor to a complete narrative and it is a bet/gamble on narrative possibilities ('ante up!'). Metanarratives influence and control how we might assemble these antenarrative fragments. They suggest the most legitimate way to narrate a story; they encourage monologue. Boje has lamented that 'so much of what passes for academic narrative analysis in organization studies seems to rely upon sequential, single-voiced renderings' (2001, p 9). But we have options for 'semantic innovation' (Ricoeur, 1984).

We can disrupt the monologue through dialogue. Boje and Smith called for the development of a dialogical approach to storytelling in entrepreneurship studies – one that uses 'multiple retrospective narrations' (2010, p 310). Elsewhere, my co-authors and I have responded to that call and restoried the University of Waterloo 'entrepreneurial ecosystem' as an entrepreneurship-producing factory, an isolating crowd, a supportive community, and a totalizing cult (MacNeil et al, 2021). The end goal of that work was deconstruction rather than composite narration (Vaara et al, 2016). Our alternate narratives were not meant to come together as a composite whole. They were incompatible with one another and worked together to problematize the idea that places (or 'ecosystems') can or should have stable and coherent entrepreneurship stories.

Similarly, the three short stories that follow work together to problematize the neoliberalization of innovation narratives. Businesses are not the main characters in this chapter. Instead, each story is centred on a different public organization: Canada's Naval Research Establishment, the BIO, and Dalhousie University's Oceanography Institute. Furthermore, market success is not the ultimate resolution in these stories; instead, the stories involve conflict and resolution around different public goods.

My narrative approach

Some readers might object that this approach to storytelling is too complex for their liking. They might prefer to consume history in one singular, modern narration. Others might accept plurivocality – the acknowledgement of several possible perspectives on the past (that is, postmodern). Either way, most readers are accustomed to one story at a time about how an industry, market, or technology emerged. And here I am asking you to read three different interconnected accounts (in addition to the three disconnected accounts I presented in Chapter 3). Like Raghu Garud and his colleagues, I am interested in the value of narrative relationalism for innovation studies (Garud et al, 2010, 2014). But while their work has focused on innovation as a narrative process, my work here is focused on innovation research as a historiographic process. Building on ideas about how we know the past (the subject of Chapter 3), this chapter considers how we might write or narrate the past differently. Like Durepos (2015), I am interested in understanding the past through narrative multiplicity – through *amodern* histories.

This is quite a different historiographic instrumentality than the default. Vaara et al have highlighted the value of poststructuralist narrative analysis 'to problematize prevailing or dominant narratives' and 'to uncover the central role of emerging narratives in organizational processes' (2016, p 15). Mol and Law have advocated a 'multi-voiced form of investigative story-telling' (2004, p 59). And Kirsch et al have written against the singular retrospective view of industry emergence. They recommend a 'deeper, contextual approach [that] reads historical evidence from the past "forward" in ways that do not foreclose alternative organizational paths' (Kirsch et al, 2014, p 229). The story would be simpler if I worked backwards from a present-day industry to construct one history. But we saw this in Chapter 2 – in the official government history, *Defined by the Sea* (Government of Nova Scotia, 2012). We have seen that looking to the past from a present-day neoliberal standpoint will always lead to stories about market-dominant evolutionary processes. This encourages public organizations to be framed as supporting characters.

Instead, I began work on this chapter with the three public research organizations that stood out during my time at the Nova Scotia Public Archives. A fourth organization, the Nova Scotia Research Foundation, also stood out in the archival records, but its mandate was economic development. Although that organization left interesting traces of ocean technology development, I have chosen to focus on the three other organizations with scientific mandates. I followed actors related to each of those organizations across many meandering traces, going back no further than the Second World War, and stopping when I reached a saturation point in my understanding of: (a) how each organization enacted ocean science; and (b) the kinds of relations each actor established with other public and private organizations. I treat each of these organizations as an actor–network (Latour, 1987; Law, 1994) and explore their efforts to produce knowledge, embed this knowledge in technology, and sometimes transfer it to other organizations. As you will see, the resulting stories do not unfold along the same timelines and do not all conclude in a common present day. I chose to begin and end each story at a juncture relevant to that organization. Although overlapping, each story is told separately. This way, there is no one universal path to the present day; each story is told for the sake of its own main character and not for the sake of understanding a present-day industry.

Three short stories

Naval research in Halifax, 1940–70

The events of the Second World War and then the Cold War changed the way the Royal Canadian Navy would know the ocean. The fledgling Navy had relied on traditional seafaring knowledge up to 1939: 'defence science still had no formal place in the activities of the Government of Canada when war broke out' (Longard, 1993, p 1). Then German aircraft started dropping magnetic mines into the Atlantic. These mines were activated by the passing magnetic field of any steel-hulled vessel. Such an unconventional weapon inspired the Navy to consult scientists. Conveniently, General Andrew McNaughton, Commander of the Canadian Forces in England, 'spoke science' (he held an MSc and had been President of the National Research Council prior to the war). McNaughton appears to have worked with the Chief of Naval Staff (Admiral Nelles) and Acting National Research Council President (Dr C. J. MacKenzie) to engage two Dalhousie University physics professors on a part-time basis in March 1940 (Longard, 1993). Drs George Henderson and John Johnstone were hired by the National Research Council and immediately seconded to the Navy. They assembled a team and began a version of the nail-wrapped-in-copper-wire experiment conducted by so many school children, albeit one where the 'nails' were ships and the goal was a near-neutral (degaussed) magnetic field. The first degaussing range in North America opened in the Bedford Basin (the interior of Halifax Harbour) in November 1940. Here, vessels could be outfitted and tested before crossing the Atlantic. This work was expanded in 1942 when a larger degaussing (and hydrophone) range opened in the main channel of the harbour. The 'Anti-Magnetic Mines Office' (later the 'Degaussing Experimental Office') performed degaussing tests on an incredible 35,000 ships in Halifax Harbour during the war (Longard, 1993, p 10). The success of degaussing earned this growing team of scientists a move from cramped Dockyard offices to HMCS *Stadacona* in 1944. They were then designated as 'Naval Research Establishment' or 'NRE' (DREA, ca. 1985, p 1). With this bump in status, the research group began to tackle new tasks.

The previous year, the Navy had called upon oceanographer Dr Harry Hachey from the Fisheries Research Board in St Andrews, NB, 'to advise on the East Coast problem' (Longard, 1993, p 51). German U-boats were lurking outside the harbour, but there were significant problems detecting the submarines with ASDIC (the British version of sonar, or 'sound navigation and ranging'). Dr Hachey did not join NRE, but helped the staff begin collecting and analysing bathythermograph (temperature versus depth) observations. The importance of these observations was heightened after the war, when German submarines were replaced by Soviet ones (Pigott, 2011). Along the way, NRE discovered that submarines were able to dive beneath a 'sound channel' of warm surface water off the Atlantic coast. Since fluctuating ocean conditions were therefore a major variable in acoustic submarine detection, the Navy needed regular oceanographic data throughout the Second World War (and the Cold War). NRE collected this data for naval operations until the BIO took over the task in 1960 (Longard, 1993, p 52). This line of research eventually led to ASDIC/sonar improvements, research partnerships with the Americans, NRE's invention of variable depth sonar (VDS), and the establishment of physical oceanography as a discipline in Canada. A breakthrough, the VDS 'towed-sonar' system (named 'CAST IX') would be built by Cossor Canada Ltd. in 1957.

While anti-submarine research was beginning in 1943, the fledgling research group in Halifax was also making a breakthrough on the problem of sea-water corrosion. All those degaussing experiments led Kenneth Barnard to develop cathodic protection. The same basic technique remains in use today by navies and commercial fleets worldwide. Barnard's innovation changed 'the whole concept of ship refits, saving untold millions of dollars' (Longard, 1993, p 41) (those interested in the technical details may refer to the patent document (US No. 3,012,959); see also Barnard, 1959).

As the war in Europe ended, NRE temporarily shrank in size. Of the 46 personnel assigned to NRE in 1945, only 11 remained in 1947. Pensions for veterans allowed many to return to school, and the degaussing range and NRE research ship were decommissioned. However, this period of 'peace time' defence research was fleeting (and research topics were unchanged). The Defence Act of 1947 then established the Defence Research Board, and NRE was reinvigorated. In five years, staffing grew to 131 people (and approximately 200 by the mid-1950s) (Longard, 1993, p 5). A new laboratory opened in 1952, and in Cold War fashion its cafeteria was reinforced as a bomb shelter (Longard, 1993, p 7). Much of the physical oceanography and acoustics research continued under an NRE department called the 'Anti-Submarine Warfare Service Projects Unit' (Longard, 1993, p 6). One of this unit's successes was the development of sonobuoy systems that proved superior to the Allies' prototypes.

Other research programmes also flourished during the 1950s and 1960s, leading to licensable patents for a variety of broad purpose technologies such as the sea-water battery (US Patent No. 4,016,399). But perhaps NRE's most ambitious cold war invention was the hydrofoil craft. Research began in 1948 with the goal of developing many small but fast anti-submarine vessels to patrol the Atlantic coast. On 24 September 1954, a photo of NRE's first working (and no longer secret) hydrofoil made the cover of *LIFE* magazine.[1] A crown corporation, De Havilland Aircraft of Canada, was contracted for design studies and to build the final 200-ton prototype. The second generation HMCS *Bras D'Or* was put to sea trials in 1967. Other published accounts tell us that the *Bras D'Or* became the world's fastest warship on 9 July 1969 (Boileau, 2004, pp 5–6). Then two years later, the Minister of National Defence announced a shift in policy from anti-submarine warfare to sovereignty protection (a focus on the Arctic). On 2 November 1971, the Minister informed Parliament of his decision to mothball the hydrofoil project (Boileau, 2004, p 82). Other work in hydronautics (naval architecture) would continue throughout the Cold War under NRE's new name, Defence Research Establishment Atlantic (DREA, ca. 1985, p 2).

Dalhousie University's Department of Oceanography, 1949–74

Dalhousie University was among three Canadian universities ambitious to start training oceanographers in the late 1940s (Mills, 1994). The Second World War had proven the science, and the Cold War was stimulating demand for the scientists (Mills, 1994, 2011; Hamblin, 2005). But funds were limited and politics were keen. In June 1949, the National Conference of Canadian Universities snubbed McGill University and recommended that the government support oceanography schools at both Dalhousie and the University of British Columbia (UBC) instead. Mills (1994) argues that UBC had already quietly secured federal support at that point and was well on its way to creating an 'Institute of Oceanography' later that summer (Mills, 1993). Meanwhile, Dalhousie was unable to make its case for financial support. Ron Hayes, Professor of Zoology, had taken the lead at Dal. He wrote to the Minister of Fisheries proposing a programme in biological oceanography, in line with his work on bacteriology (as microbiology was then known). The Deputy Minister wrote back advising that there was greater need in physical oceanography and that Dalhousie should align itself with the Joint Committee on Oceanography (JCO). The JCO included senior scientists from the Fisheries Research Board, the Royal Canadian Navy, and the National Research Council (later joined by the Hydrographic Service and the Defence Research Board). They had the political power to set Canada's oceanography agenda and were interested in the physical conditions of the ocean. UBC 'exploited the interest' (Mills, 1994, p 3). It would be a full decade before Dalhousie secured the necessary political support to offer graduate studies in oceanography on the East Coast (Mills, 1994, p 256).

However, Dalhousie eventually found success in a hybrid approach to oceanography. A partnership was announced in the 25 April 1959 issue of the journal *Nature* ('Institute of Oceanography, Dalhousie: Prof Ronald Hayes', 1959). The JCO had made an appeal to the National Research Council, resulting in a grant of $90,000 to Dalhousie. The announcement proudly proclaimed: 'All branches of marine science will come under investigation, and opportunities for work at sea will be provided by the Royal Canadian Navy, the Fisheries Research Board of Canada, and other agencies' ('Institute of Oceanography, Dalhousie: Prof Ronald Hayes', 1959, p 1161). Hayes had pulled together faculty from the departments of biology, chemistry, geology, and physics. This interdisciplinary approach was normal for the field. But the way in which it was implemented at Dalhousie created political tensions: professors were accountable to their home departments and not to the Institute (Waite, 1994).

These tensions were not to be resolved by Hayes. He left Dalhousie in 1963 to chair the Fisheries Research Board. J. E. Blanchard became Acting Director for a year while Dalhousie wooed Gordon Riley away from Yale (Waite, 1994). It has been said that Riley was 'the greatest biological oceanographer of his time' (Department of Oceanography, 2011). He became Director of the Institute in 1964 and used his political acumen to secure departmental status.

It was the University's new Life Sciences Centre that helped Riley cement[2] oceanography as a department (Waite, 1994). He chaired the building committee, while University President Henry Hicks lobbied for government support. The National Research Council offered $1 million for the facility's hallmark 'Aquatron' (Waite, 1994, pp 307–10). The tank would draw its water from the Northwest Arm of Halifax Harbour, nearly a kilometre away, allowing for 'work on the water column and on marine fish and mammals that [is] difficult to do elsewhere in one laboratory' (Department of Oceanography, 2011, p 6). The Atlantic Development Board offered another $2 million for marine biology facilities. Federal and provincial loan financing ($15 million) soon followed. Construction began in 1969 and tenants started arriving in 1971. Since the initial building funds had been earmarked for oceanographic research, this included the new *Department of Oceanography*.

Upon completion of the Aquatron, the Dalhousie student newspaper would quote Dr Kenneth Boyd's claim that it was 'perhaps the best laboratory for biological oceanographic research in the world' (Monaghan, 1974, p 1). One of the first research projects was on 'various forms of sea plankton' (Monaghan, 1974, p 1). Dalhousie oceanographers would go on to develop sensors to detect photosynthesis by phytoplankton in the ocean (Lewis and Smith, 1983) and to publish a breakthrough study in the journal *Nature* on the global decline in phytoplankton throughout the 20th century (Boyce et al, 2010). The study warns that phytoplankton are responsible for 'roughly half the planetary primary production' (Boyce et al, 2010, p 591) and their decline is evidence that 'increasing ocean warming is contributing to a restructuring of marine ecosystems' (Boyce et al, 2010, p 595).

The Bedford Institute of Oceanography, 1962–92

The Canadian Committee on Oceanography (formerly the Joint Committee for Oceanography), comprising Canada's senior government and university ocean scientists, launched a five-year plan in the early 1960s. The Committee's first priority was the construction of a government oceanography institute on the Bedford Basin of Halifax Harbour (van Steenburgh, 1962). This idea had been championed by Dr W.E. van Steenburgh, then Deputy Minister of Mines and Technical Surveys (a federal department). When announcing an initial $3 million to build the BIO, van Steenburgh's Minister 'stressed the importance of a better understanding of the oceans to science, defence, commerce, and development of the country's resources' ('Canadian Institute of Oceanography', 1959). To this list, van Steenburgh (1962) later added that the Institute would help Canada fulfil new international treaty obligations. The announcement of this 'Bedford Institute' was right on the heels of federal funding for Dalhousie University's Oceanography Institute. While van Steenburgh encouraged cooperation between the two, he publicly urged that a university scientist should 'remain free to tackle *any* problem' (van Steenburgh, 1962, p 10, emphasis in original) based on its scientific merits. Meanwhile, BIO's research programmes would be oriented to various government agendas.

This research mandate covered Atlantic and Arctic waters, where BIO would initially serve 'customers' in fisheries, navigation, maritime defence, natural resources, and weather forecasting (BIO, 1962–92, vol 1963, pp 2–4). The work was slated to include: 'Physical and chemical oceanography, air/sea and air/ice/sea interactions, marine geophysics, marine geology, tides and currents, hydrographic charting, and, in support of all these, instrument research and development' (BIO, 1962–92, vol 1965, p 5). The first scientists and staff began to arrive in July 1962: 17 employees of the Department of Mines and Technical Surveys (M&TS) and 14 employees of the Fisheries Research Board (FRB) (they were the FRB's Atlantic Oceanography Group, relocated from St Andrew's, NB). This expanded to 16 FRB and 124 M&TS employees within the first 12 months (BIO, 1962–92, vol 1962). The FRB employees retained their affiliation, setting a precedent where BIO was an umbrella facility composed of multiple government agencies. After its official opening in October 1963, scientists from the Institute's founding agencies were joined by marine geologists from the Geological Survey of Canada (1964), secretariat staff of the International Commission for the Northwest Atlantic Fisheries (1965), marine microbiologists and chemists from Environment Canada (1972), and seabird researchers from the Canadian Wildlife Service (1976). At its height, nearly 700 employees were sharing the BIO facilities and research vessels. Employees of Defence Research Establishment Atlantic, the Nova Scotia Research Foundation (NSRF), and Dalhousie University were also welcome aboard the research vessels. Originally BIO had access to five ships owned by M&TS, plus the CCGS *Labrador* (which had passed from the Defence Research Board to the Coast Guard and was used by BIO until 1977), and the CNAV *Sackville* (which remained a naval command auxiliary vessel, in service of the FRB). Some MT&S ships were built specifically for BIO use, including the CSS *Hudson* which was the first non-American and nonmilitary vessel to use satellite navigation (Clarke, Heffler et al, 2002).

Decades of research aboard these ships helped to establish Canadian sovereignty over an expanding coastal zone. Prior to the Institute's formation, a 1958 *Laws of the Sea Conference* had decided that mineral resources beneath any continental shelf should belong to the adjacent country (van Steenburgh, 1962). Then in 1977, Canada extended its fisheries jurisdiction to 200 nautical miles from shore (Nichols, 2002). Combined, these decisions dramatically expanded Canada's territory. Canada's Arctic claims were particularly contentious (Ørvik, 1982; Pigott, 2011). Although important work continued in the Atlantic, Arctic sovereignty became a critical driver of BIO research. Without question, maritime defence was the principal Arctic issue of the 1960s. BIO's second annual report explained that 'the whole ocean, from surface to bottom is or soon will be the region of potential submarine and antisubmarine operations' (BIO, 1962–92, vol 1963, p 3). But by 1970, petroleum development became the more important Arctic sovereignty issue. The search for oil and gas deposits had intensified in both the Atlantic and the Arctic. BIO had conducted some preliminary work in the northern Beaufort Sea. Then, the *Hudson '70* voyage captured global attention. BIO's *Hudson* was the first ship to circumnavigate the Americas (Nichols, 2002). While traversing the Northwest Passage, the crew conducted geological and geophysical work that 'contributed to an awakening interest in the hydrocarbon potential of this region' (Nichols, 2002, p 15). This meant that Canada's northern oceanographic concerns were not limited to Soviet submarines.

In fact, throughout the Cold War, Canada used BIO as a vehicle for Canadian–Soviet cooperation. Although BIO's closest international ties were with American institutions (such as the Woods Hole Oceanographic Institution), exchanges with Russian institutions began in 1964. That year, BIO sent one of its two Arctic oceanographers, A. E. Collin, on a Canadian delegation to the USSR (and 'the Baltic countries') that discussed 'problems of navigation in ice' (BIO, 1962–92, vol 1964, p 3). Then in 1967, BIO hosted a delegation from the USSR Ministry of Fisheries and a visit by the Russian/Ukrainian science vessel, *RN Lomonosov* (BIO, 1962–92, vol 1967–68). Ghent (1981) describes how Canada worked for years to establish knowledge flows with the USSR in Arctic Science. She notes that two memoranda of understanding were signed in 1972, including plans for further cooperation in geophysics, oceanography, and ice research.

Although working towards Soviet alliances and actively partnering with the Americans, BIO also started monitoring the Arctic for a threat posed by both nuclear powers. A programme to track marine radioactivity began in 1965. Radioactive waste was being dumped in the ocean by American, British, and Russian authorities throughout the Cold War (Hamblin, 2002, 2008). This, along with the 1970 *Arrow* oil spill off Nova Scotia, began BIO's longstanding environmental protection work (Nichols, 2002).

Clearly, the Institute was proving itself an effective *institutional* tool for the Government of Canada. But when BIO opened, most of the *technological* tools its scientists would need had yet to be invented. In fact, BIO staff started developing oceanographic instruments before their facility was ready by setting up camp in the facilities at the Woods Hole laboratory in Massachusetts (BIO, 1962–92, vol. 1962). Then, from the 1960s to the 1980s, engineers at BIO worked on a variety of ocean technologies, including 'an underwater rock-core drill, instrument mooring methods and materials, baseline acoustic positioning systems, oceanographic sensors, and seismic profilers' (Nichols, 2002, p 16). One of the BIO rock-core drills is held at the Canada Science and Technology Museum and can be viewed in their online archive. BIO employees were developing technologies like this for their own use, but sometimes also had an interest in commercialization. Many projects were developed in partnership with the private sector and then marketed internationally (Clarke, Heffler et al, 2002, p 42). One early invention, the Guildline Salinometer, was soon 'found in every oceanographic laboratory in the world' (Clarke, Lazier et al, 2002, p 25). A prototype of that device can be found at the Canadian Museum of Science and Technology in Ottawa, where it is described as:

> a major breakthrough, allowing the accurate measurement of the amount of salt in seawater; The changes in the amount of salt in ocean

water have a huge impact on climate, ocean movements and currents and marine ecosystems; The use of this instrument led to the creation of an international standard for salt measurement. (Guildline, 1973)

The partnership with Guildline Instruments Ltd. (of Smith's Falls, ON) also led to the development of the variable–depth sensor package BATFISH (BIO, 1962–92, vol. 1969–70, p 125; Watkins, 1980, p 22). The control unit for BATFISH was developed in partnership with a Nova Scotian company, Hermes Electronics, and is also held at the Canadian Museum of Science and Technology.

During the 1970s, major breakthroughs were developed with or transferred to local industry. These included: a meteorological buoy with Hermes Electronics, an ocean–bottom seismometer with the Canadian Marconi Company, and salmon aquaculture techniques that spawned[3] a billion–dollar industry (Sinclair et al, 2002). Throughout much of this time, John Brooke had led instrument development. He eventually became BIO's 'Industrial Liaison Officer'. He also sat on the Advisory Board for NSRF's Centre for Ocean Technology (NSRF, 1946–95, vols 1976–81). Upon his retirement from government in the early 1980s, he founded the company Brooke Ocean Technologies, which became a major partner for the Institute. Together, BIO and Brooke Ocean developed technologies including a 'Moving Vessel Profiler System' that improved on BATFISH, and a wave-powered profiler called SeaHorse (mentioned in Chapter 1).

But BIO's purpose here was to engage private sector resources in developing oceanographic tools, not necessarily to establish a local industry. Throughout the 1980s, additional partners/contractors outside of Nova Scotia were also involved in significant new technologies, including Huntec Ltd. of Toronto, ON (a deep-towed seismic system), Universal Systems Ltd. of Fredericton, NB (applications of the company's CARIS marine geomatics software), and International Submarine Engineering Ltd. of Port Moody, BC (the DOLPHIN and ARCS underwater autonomous vehicles). BIO's technology transfer was at a national and international scale, and it was only a means to achieving the institute's various missions.

Narrative implications

I wrote these three organizational histories with a common intent: to contravene the predominant narrative devices in innovation studies. These stories could have been combined into a unified plot. That might have given us a simpler point of departure for further research: a simple context section. But that point of departure would have contained presentist assumptions and we would find ourselves back with the problems of Chapter 3. Instead, I would like to use these short stories to 'open up' questions often hidden by narrative closure. I will argue that these three stories work together to open questions about characterization and plot. By writing this chapter differently, I get political characters and an alternative emplotment (Ricoeur, 1984).

Characterization

Public agents

In my three tales, ocean science and technology are political. We do not have dispassionate scientific rationality and objectivity (that is, the Enlightenment metanarrative). Nor do we have neoliberalism. Instead, we have a variety of other politics: Canadian sovereignty, Cold War posturing, fishing rights, nuclear waste, oil and gas exploration, and so on. The organizations are enacted through these political relations, and they organize further political relations. The sciences and technologies they produce are part and parcel of these political enactments. As Langdon Winner (1980) noted, the artefacts have politics too. The difference is that a market-based narrative abstracts the physical devices from their actor-networks, giving them 'moral and social distance' (Coeckelbergh and Reijers, 2016, p 344). Private companies are similarly given distance from politics. But this is narrative neoliberalization. Strip away neoliberal ideals and we can see that 'innovation is political all the way through' (Pfotenhauer and Juhl, 2017, p 88).

Now from this political fog, can or should we discern any essential character for public organizations? We could point to some of the archival evidence and argue that public research organizations have been the 'locus of innovation' (von Hippel, 1976) for many different technologies. We could also point to evidence that the links between science and industry were like a chain, were interactive, or were symbiotic. These stories might also suggest that the three organizations were anchor tenants (Agrawal and Cockburn, 2003; Niosi and Zhegu, 2005; Niosi and Zhegu, 2010) in a regional innovation system. Then, we might start to worry more about the systemic repercussions of the war on science. We might even conclude that 'the' innovation system is structurally dependent on these three organizations – not unlike how Silicon Valley is structurally dependent on its venture capital firms (Ferrary and Granovetter, 2009) or how Boston's biotech industry is dependent on its public research organizations (Powell et al, 2012). But notice how each of these characterizations assumes a model of innovation. And all those models are 'captive to an instrumental dyadic logic that seeks to connect technologies with markets and that sees the state as both external and subservient to those two poles' (Pfotenhauer and Juhl, 2017, p 87). Centring each public organization in its own story disrupts the reductionist and essentialized characterization of public organizations that is proposed by these models.

Characterization is unavoidable. Paul Ricoeur tells us that all narratives produce 'characters endowed with ethical qualities that make them noble or vile' (1984, p 59). This presents us with an ontological choice. Should we attempt to derive a universal role for certain characters in all stories of innovation? This would be the modernist-realist position that dominates innovation studies. Or should we work to undermine all attempts at universal characterization? This would be the postmodernist-relativist position that the mainstream abhors. The third option is to step outside the modern-postmodern debate with an amodern ontology (Durepos, 2015). Here, each characterization – every narration – is a function of a particular set of relations between authors, readers, and texts (including artefacts). The characters and narratives are entwined. Others would have done differently, but, given my motivations and the materials available to me, I wrote about public organizations that did good ('save the ocean!') and bad ('war!'). Some were even double agents (for example, 'working with the Russians!'). These organizations were active agents of innovation, and their enactments can be understood in multiplicity.

Private quartermasters

If these public organizations are the agents of innovation in their own stories (and NRE is a secret agent), then where does this leave the private companies? NRE, BIO, and Dalhousie all established complex, multidimensional relations with scientific instrumentality companies. They worked with industry to secure the necessary capital equipment and/or technical services. And so, if we are to characterize some public organizations as agents of innovation, then we might also characterize some private companies as 'quartermasters'.

In armies, the quartermaster is responsible for providing the unit with supplies. This could fit the stories in this chapter since several private companies provided important instrumentalities for the public organizations. Meanwhile, the term 'quartermaster' is used differently by navies: naval quartermasters help to navigate their ships. This could also fit these stories, since the technical expertise provided by scientific instrument companies helps to 'set the course' of research possibilities. But the army and navy versions of this quartermaster metaphor are both too simplistic for our purposes. In my stories, the boundary between public and private innovation is messier than it is in neoliberal ideology. Indeed, earlier research on scientific instrumentality innovation (von Hippel, 1976, 1988; Spital, 1979; de Solla Price, 1984; Rosenberg, 1992; Riggs and von Hippel, 1994) may have oversimplified the private sector role.

To explore this point, consider the stories of another quartermaster: the 'equipment officer' (Parker, 2005, p 4) named 'Q' in Ian Fleming's *James Bond* universe. In the early Bond films, Agent 007 is portrayed as technically inept, but highly skilled in the field (Funnell and Dodds, 2016). 'Q' and his Q Branch provide the *techne* that Bond needs to conduct his fieldwork.

Bond's knowledge of the field is his *episteme*. In many early Bond films, this contrast between *techne* and *episteme* is dichotomous, and it is embodied separately in Bond and Q (Funnell and Dodds, 2016). But in several more recent films, the division between technical and field skills is blurred – Bond becomes technically proficient and, in some storylines, Q joins him on missions – demonstrating that Q also has field skills (Funnell and Dodds, 2016; Wikipedia, 2017). In these stories, Bond and Q complement each other – not because they are opposites, but because each is somewhat proficient in the other's area of expertise. And in my stories, the various public and private organizations are entwined not only in their *episteme* and *techne*, but also in their *politika* (the affairs of their city-state).

Plot

This chapter would contain different stories if I had followed different plotlines. I could have chosen different junctures (Mills, 2010) – different points in time to begin and end my stories. I could have chosen to focus on different actors: writing short stories about a key person, technology, or business. And I certainly could have written a radically different story from the perspective of local peace or environmental activists.[4] In preparing this chapter, I took care in assembling bits and pieces of story from disparate archival records and history books. I used the historiographic tools provided to me by my academic training and networks. I made choices based on my (research) interests. And of course, I missed some important things because of these choices (more on that, including the peace and environmental activists, in Chapter 8). But we must all make choices whenever we tell a history. And those choices are always situated in sociomaterial actor–networks.

All those who study innovation are de facto historians. We write stories of the past in the form of literature reviews, context descriptions, and so on. But we do not always notice the decisions we are making about where to begin, where to end, and how to structure our stories. It is too easy to accept the most powerful narrative devices provided to us by extant theory (see Chapter 2) or during our field work (see Chapter 3). It is easy to let zombie neoliberalism take hold of our storytelling (as discussed in this chapter). This is how the field of innovation studies has ended up with so many similar stories of past innovation. We have fit our stories to the same narrative patterns. We neglect, ignore, write off, or rewrite 'abnormal' characters and plotlines. Critical historiography can break these narrative patterns and help reveal dark innovation. And that has been the key point of the past three chapters.

5

Taxonomic Classification

'I'm studying ocean tech', I would say whenever anyone asked about my PhD. That answer was conveniently short and sweet. *The Chronicle Herald* provincial newspaper kept everyone abreast of this important 'new' industry. So, on the surface, my topic was easily understood by family and friends – even those who had no idea how a PhD works. Everyone in my network seemed to understand what I meant by 'ocean technology'. But, secretly, I didn't. Something about that contrast of understandings nagged at me for months. Sure, I was caught up in the common academic problem of needing to define *everything*. But, as we started to see in Chapter 3, defining an industry is also a practical problem (Kirsch et al, 2014). And I couldn't get my head around the sociopolitics of this industrial category.

Over that first year or two, I met with several key players and policy makers to get the lay of the land (or, in this case, to 'find my sea legs'). I attended as many 'industry events' as I could. I collected names and business cards that might be useful for future fieldwork. But mostly I listened and learned. Sometimes these public events were in big conference venues (for example, see the preambles to Chapters 3 and 6). Sometimes they were the small 'ocean connector' events hosted by the Institute for Ocean Research Enterprise (IORE), a newly rebranded industry-facing unit at Dalhousie University. I followed many of the same speakers across these events, and no one was more central than IORE's Executive Director, Jim Hanlon. An electrical engineer by training, Hanlon had decades of senior leadership experience in several technology companies, including multinationals and two of his own start-ups. Much of his career has been oceans related. And anyone who followed these industry events would recognize him as a positive driving force for ocean technology development in the province. They would also regularly hear the ironically dismissive opening lines of his stump speech. He would joke that if we are going to talk about an ocean technology sector in this province, then there must only be two other technology sectors: land and aerospace.

Each time I heard Hanlon deliver this line, I would laugh along with the audience at the notion of 'land technology'. His coarse, comedic categories were clearly nothing like the animal-vegetable-mineral categories in Linnean taxonomy (Linnaeus, 1758). But regular folk do talk of aerospace technology (as do statisticians). And around these events, everyone was talking about ocean technology. On the surface, these two categories seemed viable. But they would break down whenever anyone (Hanlon included) mentioned the importance of ocean technology to space exploration. Indeed, the National Aeronautics and Space Administration (NASA) and other space agencies are heavily invested in deep-sea exploration; NASA co-chaired the international Oceans '22 engineering and technology conference. So, ocean technology is not even close to a discrete sectoral category.

Here in Canada, 'ocean tech' is the most common label for what some other parts of the world call blue, offshore, or marine tech. None of these labels is based in any system of standardized industrial classification, although I once heard rumours of lobbying efforts to gain that kind of legitimacy by 'updating' the North American Industrial Classification System (NAICS). Merely scratching the surface of these labels will reveal that they are 'folk taxonomy' – groupings only loosely connected with real, naturally material distinctions. This is akin to the way in which many people speak of spiders as 'bugs'. Biologists make clear distinctions between arachnids and insects, as they do between 'true fish' and the strikingly different species that we call starfish, shellfish, and jellyfish.

Indeed, I discovered quite a few slippery jellyfish when I tried to reproduce the list of Nova Scotia's 'over 200 companies' in ocean tech (Government of Nova Scotia, 2012, p 1). For a paper at our regional academic conference (MacNeil, 2014), I reviewed the public membership list of the Ocean Technology Council of Nova Scotia (Ocean Technology Council of Nova Scotia, 2013) (66 private sector members), the Canadian Ocean Technology Sector Map (Almada Ventures Inc., 2013) (31 private companies), a government-commissioned 'Ocean Technology' value chain analysis (Gereffi et al, 2013) (35 companies), and an internal provincial government working list (72 companies). Removing the duplicates gave me a spreadsheet of only 120 'ocean tech' companies in the province – and it included law firms, accounting firms, machine shops, and several others only vaguely linked to technology and the ocean. For 17 of these companies, I could find no trace on the internet of any product or service linked with the ocean. Nonetheless, I proceeded to assign six-digit NAICS codes to each of the firms on my list using the product descriptions on their websites and/or the NAICS codes provided in Industry Canada's (2013) directory of *Canadian Companies Capabilities*. I found that the firms were distributed across 45 different NAICS categories ranging from the 210000 level (Resource Extraction) through to the 610000 level (Education Services). This confirmed what was already clear: 'ocean technology' is pretty a 'folksy' departure from the accepted norms of industrial classification. But it also pointed towards some bigger questions about technology, innovation, and sectoral classification.

As I pulled the thread, some basic assumptions of innovation studies started to unravel. You see, product-based industrial classifications (NAICS, SIC, etc.) are embedded within innovation theory through the methods that were used by Keith Pavitt (1984) and his heirs (Archibugi, 2001; Castellacci, 2008). The resulting 'taxonomies' of innovation have been widely used as 'a predictive tool' (de Jong and Marsili, 2006, p 215) for firm and sectoral innovation performance, and this has been widely applied in public policy – most notably at the OECD (de Jong and Marsili, 2006). When I tried to map my list of ocean technology companies into these taxonomic systems, I ran into trouble (MacNeil, 2014). Four firms did not match any of the six patterns/categories identified in Castellacci's (2008) widely cited update to Pavitt (1984). For example, the website for Jasco Applied Sciences indicated that this one small company was doing two very different things: providing acoustic impact assessment services (for example, 'will this underwater activity interfere with marine mammal communication?') while also developing and manufacturing underwater acoustic sensors. The first of these matches Castellacci's 'knowledge intensive business services' innovation mode (and falls within NAICS #541712, 'Research and Development in the Physical, Engineering, and Life Sciences'), while the second matches his 'science-based manufacturing' mode (and falls within NAICS #334511, 'Search, Detection, Navigation, Guidance, Aeronautical, and Nautical System and Instrument Manufacturing'). The two modes are meant to be starkly different technological regimes, yielding starkly different technological trajectories. However, this one scientific instrumentality firm was defying the classification.

Like some Victorian-era zoologist, I became excited that I had 'discovered' the platypus – the one species that might rewrite innovation taxonomies. Biologists would label a specimen like Jasco Applied Sciences as *incertae sedis* (uncertain placement)[1] – a designation that calls for further taxonomic investigation. But then I remembered: Jasco is an organization, not an organism. It cannot be classified like an animal, vegetable, or mineral. Indeed, the entire grouping of 'ocean tech' seemed to suggest that sectoral boundaries are not as 'natural' as innovation research assumes. That is the thrust of this chapter.

As we will see, taxonomic classification is an important – and often problematic – instrumentality in the innovation studies toolkit. To understand the problems of taxonomic classification for innovation research, I build upon Gareth Morgan's (1980, 1986) work on metaphors in organization theory. Morgan (1980, 1986) argued that academic schools of thought are powerfully shaped by their acceptance and use of certain metaphors. For example, innovation studies have biological metaphors at their core (for example, Nelson and Winter, 1982) and the idea of taxonomic classification is one of those that we have explicitly borrowed from biology (see Archibugi,

2001; de Jong and Marsili, 2006). This conceptual borrowing has been theoretically fruitful, but we have long known that that there are limits to the use of biology metaphors when studying economics and organization (for example, Penrose, 1959; Nooteboom, 2000; Ziman, 2003c; Langrish, 2017). Jeroen de Jong and Orietta Marsili (2006) argue that it was important to borrow the idea of taxonomy from evolutionary biology so that we might identify the many variables that lead to differences in innovation behaviour among firms. But even Geoffrey Hodgson – the most ardent proponent of Darwinian biology metaphors in economics – notes that there are explanatory limits (Hodgson, 2002).

Taxonomic classification is what Morgan (1980, 1986) would call a 'puzzle solving activity'. Such research activities are enabled by their underlying metaphorical assumptions; the metaphors are instrumental to the research activity. In this chapter, I will argue that an implicit organism metaphor has driven the many decades of taxonomic puzzle solving in innovation studies. Unfortunately, the assumptions carried by any single social science metaphor 'are rarely made explicit and are often not appreciated, with the consequence that theorizing develops upon unquestioned grounds' (Morgan, 1980, p 619). To this end, metaphors are not merely 'literary frill' (Hodgson, 2002, p 263) or rhetorical flourish: 'the logic of metaphor has important implications for the process of theory construction' (Morgan, 1980, p 611). 'Metaphors are never innocent' (Derrida, 1978, p 17); they are instrumental.

I begin this chapter by reviewing Pavitt's formative taxonomy and its known limitations. I then use analogies from taxonomic biology to further our understanding of the three major methodological problems already described in the innovation taxonomies literature. This includes, but is not limited to, the taxonomic separation of public and private organizations. We will see how public organizations become excluded from classification. I then turn to the deeper problems that arise from the implicit use of an organizations-as-organisms analogy. This requires some general discussion of the organism metaphor and its implications for theorizing and classifying organizations. Then, I explore the theoretical assumptions of inheritance, determinism, and functional unity that are embedded in the idea of taxonomic classification. I conclude by returning to the metatheoretical implications of biological metaphors and the corresponding political assumptions that are inscribed in sectoral classification tools. I argue that we will need other metaphors if we hope to observe other dark innovation patterns.

The taxonomic puzzle

Pavitt's taxonomy

Among Keith Pavitt's countless contributions to innovation studies (Verspagen and Werker, 2004), his taxonomy (Pavitt, 1984) is said to have been the most significant (Archibugi, 2001). It sparked decades of investigations into innovation behaviour patterns and remains extremely influential to this day (Archibugi, 2001; Peneder, 2003, 2010; Castellacci, 2008, 2009). Martin (2010) goes so far as to suggest that Pavitt's taxonomy ended the race for one unifying innovation model:

> By the 1980s we had dozens, if not hundreds, of innovation studies, all coming up with their own models of the innovation process. Keith Pavitt (1984) showed that if you classify firms into a number of sectors, then you could begin to make sense of the rather baffling picture that was beginning to emerge. (Martin, 2010, p 4)

In the original paper, Pavitt used data on 2,000 'significant' manufacturing industry innovations in the UK (1945–79) to develop five broad patterns or categories of innovation behaviour. He called these his 'sectoral technological trajectories' (Pavitt, 1984, p 354). First, his 'supplier-dominated firms' were generally small, used nontechnical means to appropriate process innovations, and relied primarily on technological inputs (knowledge flows) from suppliers. His 'science-based firms' were much larger; used know-how, patents, and secrecy to appropriate a mix of product and process innovations; and drew knowledge from both internal R&D and public science. Pavitt divided his third category, 'production-intensive firms', into two subcategories: (a) 'scale-intensive firms' and (b) 'specialized suppliers'. The former were large firms that drew knowledge from both suppliers and internal R&D, and used a variety of means to appropriate process innovations. The latter category, 'specialized suppliers', were small firms that relied on their customers/users and their internal R&D activities to develop product innovations, which were appropriated through know-how and patents.

While Pavitt (1984) spoke of sectoral trajectories, some might make the distinction that he also used variables associated with 'technological regimes' (Breschi and Malerba, 1997). A technological regime is 'the particular combination of technological opportunities, appropriability of innovations, cumulativeness of technical advances and properties of the knowledge base' within which an industry operates (Breschi et al, 2000, p 388). Indeed, Castellacci (2008) describes Pavitt's taxonomy as more than simply a model of technological trajectories. The taxonomy's categories are said to help us anticipate the trajectory of a firm's technological change because the direction of that change is shaped by the firm's technological 'paradigm' (Dosi, 1982) or 'regime' (Breschi et al, 2000). Pavitt's work was an early articulation of these trajectory and regime/paradigm concepts (Castellacci, 2008).

But Pavitt's taxonomy also goes further, helping to explain the vertical linkages that tie together various manufacturing sectors. Archibugi argues that the taxonomy made it 'easier to explore how different economic units are interconnected and to identify the main knowledge flows and user-producer linkages' (2001, p 422). Castellacci has argued that these linkages are 'a crucial aspect' (2008, p 980) and the 'most original contribution of Pavitt's taxonomy' (2009, p 324). This taxonomy was therefore an early attempt to create an integrated paradigm-regime-trajectory-linkage model of organizational innovation (Castellacci, 2008).

While Pavitt's taxonomy has been used extensively by innovation scholars and policy makers, it has also been heavily critiqued (Cesaratto and Mangano, 1993; De Marchi, Napolitano, and Taccini, 1996; Archibugi, 2001; Gallouj, 2002; Hollenstein, 2003; de Jong and Marsili, 2006; Leiponen and Drejer, 2007; Castellacci, 2008). Many of the critics forget that Pavitt readily acknowledged the limitations of his work: 'Given the variety in patterns of technical change that we have observed, most generalizations are likely to be wrong, if they are based on very practical experience, however deep, or on a simple analytical model, however elegant' (Pavitt, 1984, p 370). Here, he recognized that his taxonomy was only a starting point. He called for further exploratory research, extensions, and alterations.

Gallouj (2002) provides an extensive survey of the problems with Pavitt's taxonomy. He notes that Pavitt's work excludes nonmarket firms, lumps all service sector firms into the 'supplier dominated' category, assumes that product-based sectors are heterogeneous, overlooks the possibility of innovation co-production, assumes a clear distinction between product and process innovations, neglects organizational innovations, and uses firm size as a determinant of technological trajectory (whereas the causality should likely be reversed) (Gallouj, 2002). It is generally accepted that these problems arise from methodological limitations within the taxonomic literature. Two broad remedies have been discussed: extensions and replications in other research contexts, and taxonomic research that is not linked to product-based industrial classification. Maintaining the biological metaphors for now, I call these the problems of 'mare incognitum' and 'parataxonomy'. I will explain each in turn. But first, let's consider a third unresolved problem: the 'wastebasket taxon' that Pavitt (1984) created for government organizations.

Government as wastebasket taxon

As he was concluding his original article, Pavitt noted some limitations of his work and began to suggest future alterations. In particular, he suggested that another category might be needed 'to cover purchases by government and utilities of expensive capital goods related to defence, energy, communications and transport' (Pavitt, 1984, p 370). This throwaway line tells us that Pavitt knew his exclusion of public organizations was problematic. Surprisingly, however, the taxonomic literature does not include any subsequent alterations or alternative taxonomies that integrate 'nonmarket' (public and social sector) organizations. There have been multiple attempts to develop entirely separate taxonomies of public sector innovation (Hartley, 2005; Koch and Hauknes, 2005; de Vries et al, 2016). And, of course, we know that public organizations do innovate (see de Vries et al, 2016). But generally, the public sector has been kept apart – and mostly absent – from the literature on taxonomies of innovation.

Some might say that this makes sense. They might argue that public organizations are different from private companies at the most basic taxonomic level. If so, then it would be fair to classify the whole 'public sector' (and perhaps the 'social sector') as a separate 'domain' or 'kingdom' in our innovation taxonomies. In other words, organizations that serve any public or social purpose might be analogous to bacteria and/or archaea, while for-profit companies are the eukaryotes: the animals, plants, fungi, and other familiar organisms with closed nuclei. Notice where this analogy takes us. Biology has long been dominated by the study of eukaryotic organisms. Although bacteria were first identified beginning in the 17th century – thanks to advancements in microscopic lenses, they only rose to prominence in the 19th century with the help of Louis Pasteur (among others) (Latour, 1993). Meanwhile, microbiologists only began to understand archaea as distinct from bacteria in the late 20th century. Indeed, the now common three-domain biological taxonomy – bacteria, archaea, eukaryota – was only proposed in 1990 (Woese et al, 1990). As Stefan Helmreich (2009) explains, 'alien' microbes had been found thriving in the extreme heat of deep-sea hydrothermal vents. Those microbes forced microbiologists to reconsider basic taxonomic structure and assumptions. Some say that these microbes are our ancient ancestors (hence 'archaea'). And given their love for extreme conditions, some say they might also be found on other planets. But what is most curious about archaea is that some of them have a genetic structure resembling both bacteria and eukaryota. These microbes tell us that the fundamental genetic boundaries of biological taxonomy are not as sharp as everyone assumed. This has left the tree of life 'in a brambled state' (Helmreich, 2009, p 81). I will return to the odd genetic structure of these microbes later. But for now, it should be noted that domains and kingdoms have recently been opened for major revision in biology. The biological analogy suggests that we should always be open to revising basic taxonomic distinctions.

I, for one, seriously doubt any claims to a universal distinction between market and nonmarket organizations. Yes, there are legal distinctions between publicly governed and privately owned organizations (and let's not forget the different legal structures of membership-based societies and cooperatives). But there are also so many hybrids: crown corporations, social enterprises, community interest corporations, and so on. There are all those public organizations that act 'business like' and all those private companies that purport to have a public purpose (see Rhodes, 2021). I agree with Henry Mintzberg, who once said 'it is time we recognized how limited that dichotomy really is' (Mintzberg, 1996, p 76). However, I disagree with his solution: breaking the tie with a three-domain classification of public, private, and 'plural' sectors (Mintzberg, 2015). That merely introduces a third, even more heterogeneous category. It still neglects other nonmarket innovations, like those undertaken by criminals and terrorists.

The deeper issue here is the idea of sectoral boundaries. As Patricia Bromley and John Meyer have argued, 'it is increasingly difficult to distinguish between these historically separate entities' (2014, p 939). They emphasize how difficult it now is 'to determine an organization's form (business, government, or charity) based on functional activity alone' (Bromley and Meyer, 2014, p 957). And so, I support Mark Moore's (2005) argument that we should not hold sectors as fixed. He contends that 'when we define a sector, we hold a purpose relatively constant' (Moore, 2005, p 48) and this restricts possibilities for breakthrough innovation. It is a self-fulfilling prophecy to assume that public and private organizations exhibit distinctly different innovation behaviours.

Remember, this is all moot because Pavitt's taxonomy did not include a separate category for public organizations, and nor have its successors (for example, Castellacci, 2008). Rather, Pavitt and his heirs have treated the public sector as what biological taxonomists call a *wastebasket taxon*. This is the type of shadow category that catches everything seen as taxonomically unimportant. It is a parking spot for those entities that are so uninteresting that they do not really need classification (or so uninteresting that they could be classified later by some lower-level researcher). The public organizations I described in Chapter 4 – BIO, the Naval Research Establishment, and the Dalhousie Oceanography Department – are all lumped into the same taxonomic wastebin as the motor vehicle registration office. BIO's technology development unit would only have become worthy of taxonomic classification when it spun-out as the company Brooke Ocean (to name but one example). Until that time, no one following Pavitt-style taxonomic logic would have noticed the development of technologies within BIO. The issue here is that public and private organizations are assumed to always exhibit fundamentally different innovation behaviours. Because public organizations are seen as substantially less innovative than private ones, they are seen as unimportant to the work of taxonomic classification. The careful study of public sector innovation has not placed them in a separate domain or kingdom; rather, systematic bias and lack of study have placed everything other than business in a taxonomic wastebasket.

Of the 120 'ocean tech' companies I attempted to classify in Nova Scotia, one exemplified this point. The Fundy Ocean Research Center for Energy Inc. (FORCE) is an R&D joint venture (a not-for-profit corporation) established to provide shared infrastructure facilities for the testing of four different tidal power turbines in the Bay of Fundy (site of the world's highest tides). It is an R&D facility jointly owned by the four competitor companies and their regulator (the Government of Nova Scotia). FORCE is also partnered with a variety of companies, government agencies, and research organizations (including my university). With its partners, FORCE developed a $50 million scientific sensor platform for real-time environmental monitoring, called Fundy Advanced Sensor Technology (FAST). This complex organization is not government, or business, or truly not-for-profit. It is also clearly intended to have a limited lifespan. Perhaps it will eventually become the site of a power generation facility – and then Pavitt's (1984) proposed extra category might apply. But for now, what is important is that it does not fit in our innovation taxonomies. FORCE defies the taxonomic separation of market and nonmarket organizations – of the public, private, and social sectors. It is like a mythological creature concealed in an underexplored context. This brings us to our next problem.

The dragons of mare incognitum

As we know, Pavitt's work was conducted in one specific temporal and geographical context. His taxonomy was based on data from the UK during the period 1945–79, but 'is intended to be universally applicable' (Gallouj, 2002, p 6). As Castellacci says: 'Pavitt's model ... provides a stylized and powerful description of the core set of *industrial sectors* that sustained the growth of advanced economies during the Fordist age' (2008, p 980, emphasis added). The limitation here is obvious. A large volume of research has documented the great variation in innovation patterns across temporal and geographical contexts rather than only sectoral ones (for more on this argument, see Castellacci, 2008). One test of Pavitt's taxonomy across modern-day Europe found greater variability in innovation patterns between countries than between sectors (Castellacci, 2008). The model clearly cannot be fully generalized beyond the context in which it was created. And so, if we intend to continue solving this taxonomic puzzle, we must begin to explore unusual and understudied contexts; we need what Robert Yin (2009) calls 'revelatory' cases.

This is what drove major developments in biological taxonomy. It was a thriving field in Victorian-era England. In that time and place, societal expectations of animal diversity were being upended by the kangaroos, platypuses, and other creatures being 'discovered' during colonial explorations/exploitations of places like Australia (Ritvo, 1997). Back then, European explorers had little sense of what might be found on *terra incognito*. Indeed, the unexplored and potentially dangerous places on old European maps and globes were marked with illustrations of dragons and sea monsters.

But over time, Europeans crossed the oceans and surveyed much of Earth's landmass. Their biological taxonomy became a slowly dying field. Then came reports of Archaea. *Mare incogitum* – the unknown sea[2] – captured the scientific imagination. And starting in 2000, a multidisciplinary and transnational team of 2,788 researchers undertook a 'Census of Marine Life' (Vermeulen, 2013). By 2010, those researchers had reported the discovery of 1,200 new species and had a backlog of 4,800 potential new species to be confirmed (Ausubel et al, 2010). New organisms were found in new contexts. Perhaps innovation studies could discover its own new 'species' within underexplored contexts like government or 'ocean tech'? Some of these might be as odd as FORCE.

But let's take this analogy one step further. Notice that the 'revelatory' contexts of Victorian-era taxonomy were only novel to Europeans. Before contact, hundreds of thousands of people were living in what the Europeans considered to be a hypothetical land – *terra australis*. Those Indigenous peoples were completely familiar with the local kangaroos and platypuses. As Harriet Ritvo shows, the biological taxonomy practices of 18th- and 19th-century England were situated within a culture of fear and fascination around 'hybridity and cross-breeding' – 'monstrosity and monsters' (1997, p xiii). Taxonomic classification systems and practices were entwined with Victorian England's subjugation of women (especially sexual) and racialization (even 'speciation') of anyone without white skin. As we saw in Chapter 3, context is a construct. Shifting the sociopolitical 'context' helps reveal the violence of colonization, slavery, and subjugation. In her book, Ritvo (1997) (re)worked context to show the unnaturalness of zoology and biological taxonomy. Similarly, we can (re)work the context(s) that frame innovation taxonomies.

And it is not hard to see the contextual constraints of Pavitt's postwar England. The geopolitical context is one matter. The temporal context is another. Indeed, the past 50 years have been given multiple labels – for example, 'post-industrial' (Bell, 1973), 'late capitalism' (Mandel, 1978), and 'post-modern' (Lyotard, 1984) – that all signal a departure from earlier times. But it is Jerry Davis (2022) who has most clearly articulated the implications of economic change for the classification of firms. In his book *Taming Corporate Power in the 21st Century*, he examines the problem of classifying ICT firms:

> big tech companies such as Alphabet and Amazon use their expertise in ICTs to make money. They are indifferent to industry boundaries; they look for opportunities to apply ICTs in new ways that yield profits ... They are not analogous to the railroads, oil, electricity, the telephone, radio, or a superhighway, because ICTs have become inescapable in human interaction. (Davis, 2022, p 44)

Pavitt's taxonomy was built for the era of railroads, oil, and electricity – not the era of smartphones. Countless scholars have tried to 'update' Pavitt's taxonomy, but the many adjustments and alternatives all maintain a product-based industrial structure (Archibugi, 2001; Gallouj, 2002; de Jong and Marsili, 2006). And as Davis shows us, product and process-based classifications do not work for 'unicorns' like Airbnb, Coursera, Uber, DoorDash, and Lyft (Davis, 2022). He names these companies as samples, drawn from

> a rich and diverse buffet of tech nerditude that spans hotels, restaurants, schools, highways, infrastructure construction, spying, and, apparently, clouds. And yet all these businesses are classified in SIC code 7372 ('pre-packaged software'). By tradition, this means we should regard them as competitors. Uber also classified itself as 7372, while its most obvious direct competitor, Lyft, went with 7389 ['business services, not elsewhere classified']. (Davis, 2022, p 39)

Davis is arguing that industrial classification systems are ill-suited to many of the most popularly innovative organizations in the early 21st century. He insists that the nature of these companies will contradict most fixed industrial categories. Godin (2005) made a similar point: 'biotech' does not fit into existing industrial classifications. These folk categories – biotech, infotech, ocean tech, clean tech, and so on – are rhetorical framings of different industrial contexts. We could get value from 'sensibly' mixing formal and folk classifications (Bowker and Star, 2000). But my point here is that these folk categories relate to a different industrial context than the formal ones. They reveal (and conceal) different patterns. This brings us to our third problem: innovation taxonomies are derived from Fordist-era industrial classification systems.

Industrial parataxonomy

In biological taxonomy there is a highly contentious methodological practice called parataxonomy. This approach involves sending nonspecialists into the field to collect large numbers of specimen samples and to pre-sort those samples roughly based on their most obvious physical characteristics (see Goldstein, 1997). Parataxonomy is meant to be an efficient division of labour between data collection and taxonomic analysis. However, in his critique of the method, Paul Goldstein (1997) explains that it is not necessarily more efficient – and it is certainly less effective at identifying priorities for diversity conservation. This is primarily because 'the ability of parataxonomists to sort various groups must be tested repeatedly, as must the readiness with which various groups lend themselves to sorting by amateurs' (Goldstein,

1997, p 572). Innovation studies is similar in that most taxonomic research has involved large datasets that were collected and 'pre-sorted' by others before the taxonomic researchers stepped in. Of course, I am not the first to notice this: de Jong and Marsili (2006) reviewed 11 important taxonomic contributions on organizational innovation and found that these all used some type of rough pre-sorting based on industrial product class. Indeed, a great deal of effort has been expended in innovation studies to highlight and test what I am calling the 'parataxonomy problem'.

Unfortunately, product-based industrial classifications have remained embedded in the taxonomic literature since Pavitt (1984). Pavitt established explicit correspondence between his taxonomic categories and standard product-based industries. For example, he said that his 'scale-intensive' category included firms from industries such as food product manufacturing and shipbuilding. Meanwhile, 'science-based firms are to be found in the chemical and the electronic/electrical sectors' (Pavitt, 1984, p 362). However, Pavitt had intended for his model to be a firm-level taxonomy: 'the basic unit of analysis is the innovating firm' (1984, p 353). Archibugi (2001) argues that Pavitt failed to make this clear, instead giving the impression that this is a taxonomy of industrial sectors. That impression is reinforced throughout the original paper. Pavitt (1984) performed his econometric analysis using data he had aggregated up to the industry level, defined by the UK's 'Minimum List Heading' (that is, categories of the Standard Industrial Classification). This made a great deal of sense, given his background in economic policy. But even in those times, it neglected considerable heterogeneity within each industrial class.

Since the Second World War, economists and statisticians have established a relative consensus around the product-related classification of firms. 'Standard' classification systems around the world (such as North America's NAICS, Europe's NACE, or the United Nation's ISIC) are all very similar in this respect. They sort business establishments based on the primary products they produce. However, businesses with similar product outputs can exhibit completely different innovation behaviours. A good example might be two footwear companies: one mass-producing slippers, the other collaborating with NASA to produce moon boots (Archibugi, 2001). This is the aspect of Pavitt's taxonomy that has been most heavily critiqued. For example, Archibugi (2001) has noted that multi-product, multi-technology firms defy the classification system. Gallouj (2002) has argued that the classification logic leaves no room for firms to change their product offerings or technological trajectories. Hoberg and Phillips have also lamented that none of the existing industry classifications 'reclassifies firms significantly over time as the product market evolves' (2016, p 1427). They add that the SIC, NAICS, and similar classification systems cannot 'easily accommodate innovations that create entirely new product markets' (2016, p 1427). And finally, Moore has pointed out that 'organizations can succeed by migrating from one sector to another, while a sector can improve only by getting better at producing the goods and services defined by the sector' (2005, p 48). In short, radical organizational innovation disrupts our existing industrial classifications. This means that many forms of radical innovation are excluded from data that has been pre-sorted by industry.

Cesaratto and Mangano (1993) may have been the first to statistically demonstrate that Pavitt's sector-level analysis was problematic. They argued that Pavitt's general idea was sound, but that innovation behaviours varied greatly among individual firms in Italy. De Marchi et al (1996) also tested Pavitt's model using Italian firm-level data. They found some statistical support for Pavitt's taxonomic categories, but also observed a high degree of firm-level variability within sectors. Other studies have confirmed that small firms in Switzerland (Hollenstein, 2003) and the Netherlands (de Jong and Marsili, 2006) are more diverse in their innovation patterns than Pavitt's sector-based approach might allow. Leiponen and Drejer (2007) demonstrated that firm-level capabilities and strategies were more important to innovation behaviour than sectoral characteristics in both Finland and Denmark. Considering studies like these, Archibugi argued that improvements to Pavitt's taxonomy should provide us with 'a categorization of firms entirely independent from the product-based one' (2001, p 420). He called for further innovation taxonomy research at the firm level rather than the industry level.

This is akin to Goldstein's (1997) call for biologists to spend more resources on taxonomy and to engage in less parataxonomy. Indeed, it is easy to see parallels in the policy problems that result from parataxonomy in both biology and innovation studies. Goldstein was very concerned with the serious problems biological parataxonomy can create when its results are applied in conservation policy. Parataxonomy means sacrificing the effective identification of rare and potentially endangered species in favour of more efficient rough estimates of ecosystem diversity. Similarly, de Jong and Marsili expressed concern that sector-level analysis might make for easier innovation policy, but 'it does not account for intra-industry diversity of innovation across firms' (2006, p 216). In other words, it conceals novelty within and across its categories. It encourages policy makers to apply homogeneous innovation policies to heterogeneous categories of innovators. Ironically, stimulating novelty and heterogeneity is almost always a goal of these innovation policies. We are aware that this is problematic, and yet it continues.

The organism metaphor

There have been many published adjustments and alternatives to Pavitt's taxonomy (see the discussion in Archibugi, 2001; Peneder, 2003; Castellacci,

2008). Because my focus is on the overall limitations of taxonomic logic, it is not my intention to discuss all published adjustments and alternatives. Rather, I have now reviewed the methodological challenges that persist in this literature (see also Archibugi, 2001; Gallouj, 2002; de Jong and Marsili, 2006). Many contributions to the innovation taxonomies literature end with some discussion of the various methodological limitations. These major methodological challenges – which I have summarized as the problems of wastebasket taxon, *mare incognitum*, and parataxonomy – must be addressed in future research.

But now that I have reviewed the challenges identified within the taxonomic literature, I would like to take a step back and explore the challenges that lie beneath it. I must now stop working with the biological metaphors and start working against them. In the following sections, I link taxonomic problems to three ways that organizations are not like organisms. The first of these is a reminder that organizations do not genetically 'inherit' their characteristics. Next is the agency of real organisms (humans) in the face of technological/ organizational determinism. Finally, because organizations are not actually individual entities, they do not have any natural boundaries and so they often do not have functional unity. As we will see, these three limits of the organism metaphor are boundary conditions for the taxonomic classification of organizations. But first, let's quickly review Gareth Morgan's (1980, 1986) important contributions on the organizations-as-organisms metaphor.

Organizations as organisms

Forty years ago, Gareth Morgan (1980, 1986, 1997) wrote on the paradigmatic nature of metaphors in organization studies. He argued that the relatively 'normal' (cf. Kuhn, 1962) and coherent perspectives within an academic community 'are based upon the acceptance and use of different kinds of metaphor as a foundation for inquiry' (Morgan, 1980, p 607). Any one broad paradigmatic community is often home to more than one foundational metaphor. For example, the metaphors of 'machine' and 'organism' have long dominated studies of economics (Nelson, 1995), businesses, and organizations (Morgan, 1980, 1986). Academic communities operationalize their metaphors in the various 'puzzle-solving' activities (areas and methods of inquiry) that are seen to be normal (Morgan, 1980). They are tools for solving knowledge puzzles. Metaphors thereby enable and constrain research.

Good metaphors can advance our understanding of interesting phenomena, but only to a certain extent. As Morgan explained, 'metaphor stretches the imagination in a way that can create powerful insights, but at the risk of distortion' (1997, p 5). Through metaphor we produce 'constructive falsehoods' that can be useful until they are taken literally or to an extreme (Morgan, 1997, p 5). Researchers can neglect the limits of a metaphor and, more fundamentally, can also forget that certain theories are grounded in certain metaphors (Morgan, 1980). This is how a discipline can become 'imprisoned' (Morgan, 1980, p 605) by its metaphorical assumptions. The remedy is to recognize the limitations of a metaphor. Only then can a metaphor's theorizing potential be fully realized. As Morgan said, 'in recognizing theory as metaphor, we quickly appreciate that no single theory will ever give us a perfect or all-purpose point-of-view' (1997, p 5).

In the next three sections, I am concerned with the limitations of the 'organism' metaphor for innovation studies. One might argue that innovation studies have 'inherited' the organism metaphor – and various related biological analogies – from economics. Freeman once warned that biological metaphors from economics presented 'serious dangers' (Freeman, 1991, p 211) for the study of innovation. Indeed, there is a longstanding debate about the utility of biology metaphors within economics, especially about the idea of Darwinian evolution (for example, Penrose, 1959; Nelson and Winter, 1982; Nelson, 1995; Hodgson, 2002). But I will only skim the debate about universal Darwinism (see Hodgson, 2002). The broad debate about the limits of Darwinian/biological metaphors has led to nuanced differentiation of economic 'evolution' from biological evolution (for example, Nelson and Winter, 1982; Nelson, 1995). However, the idea of 'speciation' is still contested; there is still debate around which characteristics of technology and organization can map onto a biological metaphor (for example, Ziman, 2003c; Langrish, 2017). Indeed, the neo-Darwinians who study innovation know that organizations are not exactly like organisms. But they nonetheless write of 'taxonomic classification' – and this carries the organism metaphor forward. Thus far, the puzzle-solving activity of taxonomic classification has been beyond the scope of the biological metaphor debate. My focus is therefore not on 'evolutionary' processes in economics per se; it is more tightly constrained to certain 'taxonomic' implications that arise from the organization-as-organism metaphor.

The 'organism' metaphor works because it allows us to treat organizations as relatively stable entities that exist within an environmental context. Morgan explains that

> in the organismic metaphor the concept of organization is as a living entity in constant flux and change, interacting with its environment in an attempt to satisfy its needs. The relationship between organization and environment has stressed that certain kinds of organizations are better able to survive in some environments than others. (Morgan, 1980, pp 614–15)

This has led to some of the most important theoretical insights on business and organization (Morgan, 1986). For example, it is the metaphor that inspired systems theories and life-cycle theories of the firm (Morgan, 1986, 1997). It is a core assumption of the 'population ecology' approach to organization theory (Morgan, 1980, 1986), which brought Darwinian ideas into organization studies (for example, Hannan and Freeman, 1977). It is at the centre of the sociotechnical systems approach (Morgan, 1986), which examined the mutually constitutive nature of organizing and technology (for example, Trist and Bamforth, 1951). And in the context of this chapter, it is particularly important to note that the organism metaphor is the foundation for contingency theory (Morgan, 1980, 1986), which theorized different forms of organization as adaptions to different environmental conditions (for example, Burns and Stalker, 1961). It was early contingency theorists who noticed how different forms of organization were linked with different technologies (for example, Burns and Stalker, 1961; Woodward, 1965). This, of course, led to decades of research into the various 'species' of organization (Morgan, 1997). However, all this theorizing has found its limits. Following Morgan (1980, 1986), I argue that those limits exist because organizations are not biological organisms. In fact, what these theoretical traditions all share is that they reify the processes of organizing; they metaphorically create a real and concrete entity (an 'organization') out of sociomaterial interactions.

'Inherited' characteristics

Throughout the debates on biology metaphors in economics, the most hotly contested question has been whether organizations 'inherit' their traits. And when it comes to innovation taxonomies, a central assumption 'is that firm behaviour is shaped and constrained by the nature of the technologies they use' (de Jong and Marsili, 2006, p 214). In other words, organizations inherit their innovation behaviours through their technologies. But how far does this 'inheritance' analogy (and, by extension, the evolution analogy) stretch? What are its limits? In building his argument for universal Darwinism, Hodgson admits that 'the strongest reasons to be skeptical of "biological analogies" involves the detailed differences between the types of evolutionary mechanism applying to the socio-economic and to the natural domain' (2002, p 274). He argues that the concept of inheritance (and replication) is where universal Darwinism will 'find its boundary' (Hodgson, 2002, p 273). In this section, I argue that the idea of inherited traits takes us beyond the useful bounds of the organism metaphor.

To rethink the ways in which organisms and organizations are different, consider one of biology's ongoing taxonomic problems: the classification of larvae. For a period in the early 19th century, 'zoea' were considered a type of crab (a genus). Then taxonomists realized that 'zoea' was a larval stage in crab development. It is sometimes difficult to distinguish between larvae and adult animals because classic taxonomy relies on the identification of unique physical characteristics. Of course, these can change dramatically as the research subjects mature. When ocean scientists are searching for new species, the samples returned in their nets can often be difficult to sift through, and trained taxonomists tend to have large backlogs of work. In response, many scientists turned towards DNA barcoding (Snelgrove, 2010), which is not only faster but can also work around the maturity problem (although, as we will see later, even this has its limits). Generally, larvae and adults of the same species contain the same genetic material. Indeed, that genetic material is what determines the stages of an organism's life cycle and its physical features at each stage. Although there have been attempts to develop life-cycle taxonomies of firms (for example, Abernathy and Utterback, 1978), it is difficult to argue for material determinism in organizational life. As Freeman (1991) asserted, the life cycle of firms is not like the life cycle of organisms. There is no DNA barcoding solution for the study of organizations.

And there is no debate on this point: organizations do not have DNA. Nelson and Winter may have argued that, in economics, 'routines play the role that genes play in biological evolutionary theory' (1982, p 14), but they were aware that this was an analogy. Nonetheless, Hodgson (2002) took their argument one step beyond metaphor: 'One possible and relevant example is the propensity of human beings to communicate, conform and imitate, making the replication or inheritance of customs, routines, habits and ideas a key feature of human socio-economic systems' (Hodgson, 2002, p 270). Based on this kind of reasoning, some scholars have come to accept 'memes' as a corollary to 'genes' within evolutionary economics. But Ziman (2003a) characterizes the move towards 'meme' as merely a 'convenient' way to 'sustain the overall analogy' (2003a, p 5). Witt (1996) has forcefully argued that there is nothing resembling genetic material at the social level. Louçã and Cabral (2021) have asserted that 'no economic analogue exists for the replication unit in biology' (2021, p 4). Patterns of human behaviour – like organizational routines – are simply not analogous to genes; rather, because humans enact organizations, their genetic material is only part of any organizational process (Vromen, 2006).

Organizing involves social interaction between humans. It also involves layers of social behaviour between other constituent materials, such as technological artefacts. Since the earliest days of the organization-as-organism metaphor, we have known that social and technological realities are mutually constitutive (Trist and Bamforth, 1951; Burns and Stalker, 1961; Woodward, 1965). We know that organizing is a sociomaterial process. But the organism metaphor still implies that organizations are materially determined – in the same way that genetic material determines the make-up of an organism. Indeed, by 2004, Nelson was recommending that we abandon 'the search for credible analogues of genes in economic processes' (cited in Vromen, 2006, p 544). There is no definitive DNA barcode that we can sample; firms do not have fixed, technically determined characteristics.

But this has not stopped us from overextending the analogy of inherited characteristics. In taxonomic logic, organizations are seen as being 'locked' into a particular sectoral trajectory (Gallouj, 2002). Or, as de Jong and Marsili (2006) explain, technological regimes determine 'the directions, or 'natural trajectories', along which incremental innovations take place within the regime' (de Jong and Marsili, 2006, p 215). There is most certainly a path dependency to innovation behaviours (Cohen and Levinthal, 1994). But the organization-as-organism metaphor neglects the very important role of human agency (Morgan, 1986). Morgan (1986) notes that this is one of the major limitations of the organism metaphor: it easily gives way to an ideology of natural determinism and Social Darwinism.

Innovative agency

I join others in arguing that the organism metaphor leads us towards an overly deterministic perspective on innovation and organization(s). These arguments go back at least as far as Edith Penrose (1959), who said that we should reject theories that treat firms as if they are organisms because such theories neglect human agency (see also Levallois, 2011). She said 'to abandon their [firms'] development to the laws of nature diverts attention from the importance of human decisions and motives, and from problems of ethics and public policy, and surrounds the whole question of the growth of the firm with an aura of "naturalness" and even inevitability' (Penrose, 1959, p 809). Chris Freeman applauded her argument and applied it to innovation theory. He agreed that we should not 'give explanations of human affairs that do not depend on human motives' (Freeman, 1991, p 219). I might add the agency of other actors to the entanglements we call organizations/ organizing (MacNeil and Mills, 2015). But then we simply get a more complex and distributed sociomaterial agency. Nonetheless, Penrose was right to assert the agentic power of human managers and thereby confront the conservative bias (Levallois, 2011) that we find entwined with the organism metaphor.

Contingency theories of the firm developed around this sense of agency. These theories focus on how managers can adapt their strategies (and organizational forms) to their environments. But some theorists argued that this allocated 'too much flexibility and power to the organization and too little to the environment' (Morgan, 1997, p 60). This led to the more Darwinian 'population ecology' approach to organizations (see Hannan and Freeman, 1977). Although it produced some valuable insights, many organization theorists see population ecology as far too deterministic (Morgan, 1997). Through an overextension of the organism metaphor, the population ecology view can take the 'rather pessimistic stance that this choice [human agency] will never count for much because environmental forces ultimately have the upper hand' (Morgan, 1997, p 68). This downplays or discounts the role of human creativity in organizing and the ability for organizations 'to create market niches that never existed before' (Morgan, 1997, p 63). In short, it discounts many possibilities for innovation.

We cannot afford to take the organism metaphor this far in innovation studies. As Vromen (2006) suggests, 'the irregular, unpredictable parts in firm behaviour might be considerable' (p 559). Indeed, my view is that those irregular, unpredictable parts should be a primary focus of innovation studies. But the organism metaphor focuses us deterministically on inherited traits and the ways that organizations conform to their contexts. There are certainly *material* aspects of organization. But unlike organisms, organizations are also *socially* constructed (Morgan, 1986).[3] This is the greatest limitation of the organism metaphor: 'Organizations are very much products of visions, ideas, norms, and beliefs, so their shape and structure is much more fragile and tentative than the material structure of an organism' (Morgan, 1997, p 69). The social shaping of technology (MacKenzie and Wajcman, 1999) is taken for granted in science and technology studies. However, innovation taxonomies are much more deterministic.

Taxonomies of innovation assume that organizations exhibit patterns of behaviour that are determined by – or inherited from – their technical reality. This assumption is what allows us to place organizations into mutually exclusive categories. But it only works for organizations that have stable technological regimes/trajectories. They must have characteristics that are determined as if by some genetic/inheritance mechanism. An organism cannot choose to change its phylogenic characteristics, but organizational decision making can change innovation behaviours. As I have already noted in this chapter, some organizations undertake radical change. In fact, some organizations are known to completely redefine their domains (Covin and Miles, 1999). Taxonomic logic could be applied to these organizations, but they would need to be awkwardly classified as different species before and after any radical change (much like our crab larvae, zoea). I therefore conclude that the boundaries of the organism metaphor do not extend to radically innovative organizations that are in the process of establishing new sociotechnical realities. This metaphor leaves those organizations and innovations in the dark. Again, we see that taxonomic classification is unsuited for observing some of the most interesting innovation phenomena.

Functional (dis)unity

The organism metaphor's other major boundary condition is interesting because it relates directly to the 'real' boundaries of organizations. Treating organizations as organisms allows us to reify clear limits around our unit of analysis: to treat organizations as open systems. Morgan explains this part of the analogy:

> If we look at organisms in the natural world we find them characterized by a functional interdependence where every element of the system under normal circumstances works for all the other elements. Thus, in the human body the blood, heart, lungs, arms, and legs normally work together to preserve the homeostatic functioning of the whole. The system is unified and shares a common life and a common future. (Morgan, 1997, p 70)

However, organizational processes are not unified in this way. Organizations do not have natural boundaries (except sometimes when they operate in a single building). And so, they seldom have functional unity. At the extreme, some criminals and terrorists find innovative ways to avoid functional unity in their organizing.

Elsewhere, Albert Mills and I have shown that the idea of an organization is a black box, and the processes of organizing are much more precarious than we typically realize (MacNeil and Mills, 2015). The organism metaphor helps us to temporarily suspend the messy processual realities of organizing and thereby study discrete, stable 'entities'. To put it another way, 'organizations are but temporary reifications, because organizing never ceases' (Czarniawska, 2004, p 780). It is often empirically useful to temporarily 'pin down' organizational life like an entomologist might pin down an insect. The problem is that the organism metaphor determines our unit of analysis. As Morgan says, organizations 'are not discrete entities, even though it may be convenient to think of them as such' (1997, p 64). Ontologically, organizations are not really (materially) *individuals*.

This has manifest as a messy methodological problem for taxonomies of innovation. Although they followed the normal logic of the organism metaphor, de Jong and Marsili (2006) noticed this problem. In their final sentence they say: 'A suggestion for future work is to integrate the different levels of analysis and disentangle the influence that conditions specific to single innovations and to industrial sectors have on the diverse clusters of innovative firms' (de Jong and Marsili, 2006, p 227). And so, it has been acknowledged that sometimes 'innovations' are embedded within 'organizations' which are embedded within 'sectors'. But with so many possible units of analysis, which is the right one? Do we go 'down' to the level of the technique or routine, as argued by Joel Mokyr (2013)? Or do we follow Nick Oliver and Michelle Blakeborough (1998) towards the distribution of innovation across networks of multiple organizations? If we had only these two choices, I would be inclined towards the latter – because Oliver and Blakeborough (1998) included two scientific instrument innovations among the seven aggregated within their study. However, when it comes to taxonomic classification, the organism metaphor tells us what to do: we treat organizations as the embedded unit of analysis within industrial sectors. Beyond the metaphor, we know that innovation processes are not bounded within organizations in the way that biological processes are seen to be bounded within organisms.

To understand this limitation, let us turn to the 'coral reef problem' from early biological taxonomy. (Please hold this example loosely; I am merely turning the organism metaphor in on itself, before tossing it aside.) The 'rocks' that most people consider to be coral are actually the limestone exoskeletons of many individual polyps, smaller than your littlest finger, living in densely packed colonies. These polyps also live symbiotically with algae, sometimes inside their bodies. Turning the original Linneaen taxonomy on its head, corals are therefore part animal, part plant, and part mineral. Today it is much easier for a taxonomic biologist to distinguish between each part of a coral. Algae living within a polyp can be identified as separate and distinct organisms through genetic barcoding. But two major difficulties arise when we try to do the same for discrete, mutually exclusive types of organization: the hybridity problem and the symbiosis problem.

Hybridity

First, we struggle to classify hybrid organizations. De Jong and Marsili (2006) explain this as a methodological limitation of taxonomic research: 'the same firm may implement various types of innovations ... each single innovation may display different patterns' (2006, p 227). However, shifting the unit of analysis to a smaller (or larger) scale does not guarantee that we will find functional unity. That would be akin to searching for algae inside a polyp. But with organizing we will never find DNA. It makes more sense to address this problem by shifting away from the organism metaphor entirely, since 'unlike in nature, where species are distinguished by discrete clusters of attributes, organizational characteristics are often distributed in a more continuous way. One form often tends to blend with another, producing organizations that have hybrid characteristics' (Morgan, 1997, p 55).

This hybridity is possible at all sizes and scales of organizing. Neglecting it is extremely problematic for innovation studies because if we follow Schumpeter's (1934) logic – if innovation is, indeed, the implementation of new combinations – then innovation processes will always produce hybrids that defy classification. Certainly, Ritvo (1997) has shown that biological taxonomy developed through cultural opposition to hybridity. Hybrid animals like the platypus 'were stigmatized ... not only as mongrels but as "monsters"' (Ritvo, 1997, p 132). She quotes the famed English wood-engraver and naturalist Thomas Beswick to illustrate some of the beliefs that informed biological taxonomy: 'Nature has providently stopped the ... propagation of these heterogeneous productions, to preserve, uncontaminated, the form of each animal; without which regulation, the races would in a short time be mixed with each other, and every creature, losing its original perfection, would rapidly degenerate' (Beswick, cited in Ritvo, 1997, p 89). Presumably, no innovation scholar today would ascribe to such a position. Hardly any would be aware of the racist context surrounding early taxonomic practices. But the fact remains that taxonomic classification in biology was imbued with the values of purity and pedigree. Taxonomic classification will always be at odds with hybrid entities.

Symbiosis

Attempts to identify discrete, mutually exclusive types of organization can also miss symbiotic, mutually dependent organizations. Again, this has been framed as a methodological limitation for innovation taxonomies. For example, Gallouj (2002) criticized Pavitt's taxonomy for missing the possibilities of innovation coproduction between service providers and their customers. But I argue that the limitation here is found in our metaphor, not our analytical methods. Consider the food chain and food web metaphors for a moment (again, for brief rhetorical effect). When scientists were working within the 'food chain' paradigm, marine bacteria did not seem to be very important. The linear chain started with phytoplankton performing photosynthesis and proceeded through progressively larger marine animals. Then better techniques allowed biologists to understand the important role of various bacteria in digesting dissolved organic matter. These bacteria are consumed by single-cell zooplankton, some of which are taken up by larger organisms, but many of which cycle back into dissolved organic matter (and become food for the bacteria). These previously unnoticed microbial loops form the foundation of the marine food web. It took a switch from the food 'chain' metaphor to a food 'web' metaphor to recognize the importance of this marine subsystem.

The logic of existing innovation taxonomies predicts vertical chains between organizations in different categories. This is based on the linear thinking of inputs and outputs. But we know that there are interesting innovation behaviours in this world that depend on the strategic mutual dependence of two or more different organizations. For example, consider Mowery's work on innovation within the military-industrial complex (for example, Mowery,

2009). As Mowery and Nelson (1996) have said, 'corporations are one part of a complex institutional system, and their role cannot be understood in isolation' (p 189). Also consider Gorm Hansen's (2011) research on the symbiotic relationship between a biology research laboratory and a biotech firm. These two organizations were innovating as one – a phenomenon that is invisible to innovation taxonomies. The organism metaphor fails to classify these forms of organizing because the innovation processes do not have functional unity within a concrete boundary.

Moving beyond speciation

Away from biological metaphors

This chapter has considered the challenges of taxonomic classification in innovation studies. Many scholars have attempted to solve the taxonomic puzzle and their insights have had a considerable positive impact on innovation theory and policy (de Jong and Marsili, 2006). Overall, taxonomic classification has helped us to reduce empirical complexity by establishing 'few and easy to remember categories' (de Jong and Marsili, 2006, p 214). From Pavitt (1984) onwards, the resulting taxonomies have guided researchers and policy makers in finding, recognizing, and reinforcing innovation behaviours. But as we have seen, these taxonomic classification tools have clear methodological and ontological limitations. They conceal many interesting forms of innovation.

We could reveal some dark innovation by improving these taxonomic tools. For example, we could drop the practice of industrial parataxonomy. But, as we have seen in this chapter, some of the taxonomic issues are inseparable from the biological analogy that enables this whole exercise. The organism metaphor is useful in many ways, but all metaphors have limits (Morgan, 1980, 1986). As Morgan said, 'any one metaphorical insight provides but a partial and one-sided view of the phenomenon to which it is applied' (1980, p 611). The challenge is that individual metaphors can powerfully shape social scientific paradigms (Morgan, 1980, 1986).

In the 1990s, a collaboration of British researchers spent four years considering the extent to which biological metaphors should be applied to technological innovation. This "Epistemology Group" could not establish a consensus on either maintaining or abandoning analogies to biological evolution (Ziman, 2003c), but they did conclude that we should reject the idea of speciation: 'we no longer feel impelled to find technological analogies for the most familiar evolutionary concept in biology – the notion of a *species*' (Ziman, 2003c, p 313, emphasis in original). John Ziman was lead author of the book that arose from this collaboration, and his chapters were the only ones to directly confront taxonomic classification (albeit briefly). In his opening chapter, Ziman argued that we should 'give up such Procrustean exercises as trying to make industrial firms look just like organisms' (2003a, p 11). And in a later chapter he bemoaned that 'taxonomies that mirror "the tree of life"' are 'all too tempting' (Ziman, 2003b, p 46). Unfortunately, many innovation researchers still succumb to the temptation.

Perhaps this is because taxonomic classification is such a unifying tool for such a loosely defined field of study. Although Chris Freeman 'voiced strong reservations about the uncritical translation of biological concepts' throughout his career (Louçã and Cabral, 2021, p 2), he once said that 'a taxonomy is essential both for analytical purposes and as a tool for empirical research [in innovation studies]' (Freeman, 1991, p 222). He knew that this tool both helped and hindered innovation research. Indeed, he argued that 'any such taxonomy or classification system must of course do some violence to the infinite complexity of the real processes of technical and economic change' (Freeman, 1991, p 222). I have concurred with him on this point. But my arguments in this chapter have diverged from Freeman's very early assertion that 'all schemes of classification are to some extent arbitrary and artificial' (Freeman, 1974, p 261). Innovation taxonomies are not arbitrary. By exploring the metaphors that enable this exercise, I have surfaced some of the assumptions carried from past sociopolitical contexts. Pavitt's taxonomy was key in the formation of innovation studies as a field (Fagerberg and Verspagen, 2009). But it was also, perhaps unintentionally, one inscription point for conservative, neoliberal ideas about innovation.

In their book *Sorting Things out: Classification and Its Consequences*, Geoffrey Bowker and Susan Leigh Star explain that 'standards and classifications, however dry and formal on the surfaces, are suffused with traces of political and social work' (2000, p 49). And we have now seen some of the social and political traces that shaped taxonomies of innovation. Nonmarket organizations were set aside. Assumptions about industrial structure were anchored in postwar England. Standardized industry categories were adopted for methodological efficiency. These processes, which I discussed in the first half of the chapter, have been somewhat easy to identify in the taxonomic literature. However, the literature treats these decisions as apolitical. Of course, they cannot be. Bowker and Star note that 'each standard and each category valorizes some point of view and silences another. This is not inherently a bad thing – indeed, it is inescapable' (2000, p 5). The bad thing is to deny these politics.

We have 'consistently ignored' the insidious conservative bias attached to biological analogies (Levallois, 2011). Harriet Ritvo (1997) showed us how biological taxonomy became imbued with Victorian English values. Taxonomic practices developed in favour of purity and against hybridity. And Edith Penrose could see the conservative bias in her time as well – a time when Social Darwinism was resurging and her friends were being persecuted under

McCarthyism (Levallois, 2011). In addition to the material harm caused to people with progressive views (such as McCarthy's 'communist' witches), Penrose could see the harm to organizational theory: 'the chief danger of carrying sweeping analogies very far is that the problems they are designed to illuminate become framed in such a special way that significant matters are frequently inadvertently obscured' (Penrose, 1952, p 804). At various points in this chapter, we have seen that the combined organism-species-taxonomy analogy obscures many forms of innovation. There is the neoliberal effect where firms are framed as the primary organisms for innovation. But there is also the obscuring of heterogeneity, divergence, hybridity, and symbiosis – processes that ought to be celebrated in innovation research.

In Chapter 2, I argued that theoretical models are social scientific instruments. I said that even when models are tacit, we can think of them as noncorporeal actants (Hartt, 2019). They work through physical traces, and through our sensemaking, to co-construct knowledge. Now, we can add theory-laden metaphors to that mix. The organization-as-organism metaphor is a noncorporeal actant in our taxonomies of innovation. It is entwined in a sociomaterial knot of taxonomic instrumentalities. This enables and constrains our understanding of innovation. Once we start looking, we find bits and pieces of biological analogy everywhere in the taxonomic toolkit. As Bowker and Star state, 'all classification and standardization schemes are a mixture of physical entities, such as paper forms, plugs, or software instructions encoded in silicon, and conventional arrangements such as speed and rhythm, dimension, and how specifications are implemented' (2000, p 39). In other words, classification systems appear highly structured, but they are a messy sociomaterial stew.

Towards other classification tools

Because the organism metaphor is taken for granted, my arguments here are likely to encounter considerable resistance: 'Schools of theorists committed to particular approaches and concepts often view alternative perspectives as misguided, or as presenting threats to the nature of their basic endeavour' (Morgan, 1980, p 613). For those who might object in this way – those who are wedded to some adaptation of the biological analogies – let me point out that biological taxonomy is undergoing a similarly uncomfortable paradigm shift. You see, those microbes that were found in the deep ocean – the ones we now call Archaea – have been found to actively swap genetic material. They cannot be classified based on their DNA. Helmreich explains that this means 'the stability of the category of *species* for microbes has been called into question' (2009, p 87). He suggests that 'the tree of life might turn out to be a net' (Helmreich, 2009, p 82) and that biology might be headed in a direction that 'strikes a familiar chord with readers of the maniac philosophy of Gilles Deleuze and Félix Guattari' (Helmreich, 2009, p 83). Wherever biology might be headed, it too must grapple with the design of its taxonomic tools. Too many organisms in the deep sea do not fit within a classification system that carries assumptions from the land. Something similar is true for innovation studies, where there are known instances of innovation that do not fit within old, British, industrial assumptions. And there are most certainly 'known unknowns' – readily identifiable instances of hybridity, symbiosis, and other kinds of dark innovation – that cannot be directly observed using the organism metaphor.

Interestingly, innovation studies might be primed for a shift. Ziman (2003a) has already suggested that technological innovation might be better classified using a 'neural net' metaphor. Meanwhile, Hoberg and Phillips (2016) have developed an approach to classifying publicly traded firms in the US based on the interrelatedness of text in their public disclosures – an approach they call 'textual network industry classification'. But networks (and rhizomes) are not the only possible alternative classification devices. At one time, before the tree of life became locked in, there were other unusual zoological classification systems, such as the 'quinary' approach of grouping organisms around interlocking circles (Ritvo, 1997). In the next chapter, I look to Law and Singleton (2005), who considered four metaphors – region, network, fluid, and fire – for theory building in science and technology studies. But regardless of the specific metaphors we choose, let's heed Morgan's advice: 'in order to understand any organizational phenomenon many different metaphorical insights may need to be brought into play' (Morgan, 1980, p 613). The overuse of this one metaphor has focused research on those innovation phenomena that fit. Other metaphors will help illuminate dark innovation.

The organism-species-taxonomy exercise has found its limits. This does not mean that classification should cease; rather, it means we should explore new metaphors 'which overcome the weaknesses and blindspots of traditional metaphors, offering supplementary or even contradictory approaches' (Morgan, 1980, p 612). I take up the challenge of new metaphors in the next chapter.

6

Surveying Topologies

It was too early on a Monday morning to make sense of the massive machine in front of me. The tour guide introduced it as North America's largest geotechnical centrifuge. But it looked like a poorly designed carnival ride: two boxes, each big enough to hold a human, on either end of a massive rotating arm. The official size: 200G. Not grams, but 'g' forces (I had to look it up). This thing can produce forces equal to 200 times Earth's gravity. And there it was, sitting in a cement bunker at Memorial University – on the edge of the Atlantic Ocean and as far east as one can get while still being in the Americas. I was confused: why here? What for? (And where's the coffee?)

I had come to the city of St John's, in the province of Newfoundland and Labrador, to attend *Oceans '14* – an international ocean technologies conference. This tour of the Memorial University facilities was a warm-up event for the highly technical conference sessions. It had been six years since the IEEE Oceanic Engineering Society and the Marine Technology Society had convened their joint annual conference here in Canada. So, this was a rare chance to get a 'broader', 'global' sense of ocean science and technology. It was also a chance to see ocean technology at play in a different province. (Never mind that my wife, toddler son, and newborn daughter could tag along to visit relatives.)

Some people had told me that ocean technology was 'different' in Newfoundland than in Nova Scotia. And the differences began to resonate in those first hours of the conference. I was disoriented by a deluge of engineering talk as our tour group walked through Memorial's facilities. At home, my tours of university and public research facilities were almost exclusively focused on the work of ocean scientists. Their gizmos and gadgets – even the massive 'Aquatron' research tank at Dalhousie – were props for stories of scientific intrigue. But at Memorial, the stories were about the technologies themselves. The centrifuge was one thing: it is used to model interactions between sea ice and the land. Elsewhere at the C-CORE lab, my tour group heard about the development of technologies to detect icebergs from space and to protect vessels/oil rigs from iceberg impacts. At the Autonomous Ocean Systems Lab, we heard about the development of technologies to propel and control underwater autonomous vehicles. These vehicles might carry scientific instruments and enable new oceanic observations. Indeed, we heard a bit about the amazing deep-sea cameras developed by my wife's high school friend, Dr Adam Gobi. He was the lead engineer on the camera systems that James Cameron took into the Mariana Trench. Cameron's film *Deep Sea Challenge* had been released only a month before the conference. But even here, when the facility tour turned to questions of exploration, the emphasis was on the engineering feats – the advancements in *techne*. New scientific discoveries – *epistemic* advancements – were mere ellipses at the end of each story.

I left Newfoundland convinced that ocean technology was very different there than at home in Nova Scotia. One place privileged ocean engineering and the other privileged marine biology. And yet, there were people in my ear suggesting that I should merge these into one region – that I should include all four Atlantic Provinces in my study. This would match the Government of Canada's approach: its Atlantic Canada Opportunities Agency has a four-province mandate. I could find very little in the way of ocean technologies in the other two of those provinces. But apparently federal officials wanted to encourage inter-provincial collaboration and thereby reduce resource competition between Memorial University (in Newfoundland) and Dalhousie University (in Nova Scotia). There was one speaker at the conference whose talk extended the boundaries even further. Dr Doug Wallace, the Canada Excellence Research Chair in Oceanography at Dalhousie University, lumped all of Eastern Canada – the 2,500 km drive (and ferry ride) from Montréal to St John's – into one 'ocean observation technology cluster'. He also spoke about promising tripartite discussions between Canada, the US, and the European Union that might lead to a collaborative scientific observation system for the 'whole' North Atlantic Ocean. I liked his ideas. But frankly, I just needed a clear sense of where to start and stop my PhD research. All this talk of shifting boundaries, and the 'global' nature of this conference, made that problematic.

In this chapter, I confront the problems of boundary specification for innovation studies. I begin by considering how *regional* or *volumetric* thinking – one form of spatial reasoning – drives the systems of innovation literature right into a theoretical brick wall. That wall is made of the surveyable, measurable, *Euclidean* spaces we call innovation systems. The wall is buttressed by a static form of institutional theory; institutional fields are conceived as containers for interactive learning networks/relationships. But this makes it difficult to observe innovation processes that fold people, things, and places together in new ways. In the first parts of this chapter, I discuss the ideas underpinning this approach to space, place, and innovation. I then describe how those ideas bounded my survey of innovation in ocean science instrumentalities. All this requires epistemologically positivist and ontologically realist assumptions. We will see that this form of thinking allows us to observe only certain kinds of innovation, in defined fields of technology, and in particular places.

In the conclusion to this chapter, I explore what might be missing from the regional/volumetric view. I argue the need for other forms of spatial reasoning in innovation studies. I take 'object lessons' from John Law and Vicky Singleton, and briefly consider the different implications of framing innovation as a region, network, fire, or fluid object (Law and Singleton, 2005). To understand the limits of thinking with region and network metaphors, I share 'excess' data that flooded and set fire to the boundaries I had established for my survey of scientific instrumentality innovation. In this chapter I ask: what observations about innovation are left in the dark by an over-reliance on one or two spatial metaphors?

Regional networks

Innovation system boundaries

Let's begin with the dominant spatial perspective: regional networks. Much innovation research works from the premise that mappable regions contain networks of interactive learning. This is almost taken for granted today. Of course, regional approaches to innovation have a long history. We could go back to the knowledge that was suspended 'in-the-air' of Alfred Marshall's (1890) industrial districts. Or Chris Freeman (1995) would have us go back further and notice the (possible) influence of Frederich List's *National System of Political Economy* (1841). But hardly anyone today has read those works. The ideas they contain are dead ancestors to the *modern* systems of innovation. And please note the double entendre in my use of the word *modern*. Many scholars would argue that innovation systems are the *most modern* theoretical framework. But in this chapter I will contend that the systems of innovation approach is an outdated exercise in social science *modernism*.

In Chapter 2, we saw the beginnings of the innovation systems approach in Christopher Freeman's writing about Project SAPPHO. In that preincarnation, the UK was an implicit boundary within which Freeman could think about the systemic nature of innovation. Freeman brought geopolitical boundaries to the forefront in his 1987 book. There, he explained how unique institutional arrangements, such as industrial groups or *keiretsu*, developed into an effective 'national innovation system' (NIS) for postwar Japan (Freeman, 1987). Over time, innovation systems would take on a variety of subnational and supranational boundaries. Charles Edquist (1997), who appreciated the value of vague boundaries, emphasized the many overlapping possibilities: 'innovation systems may be supranational, national or subnational (regional, local) – and at the same time they may be sectoral within any of these geographic demarcations. There are many possible permutations. Whether a system of innovation should be spatially or sectorally delimited depends on the object of study' (Edquist, 1997, p 12). And so, there is a wide menu of boundary options in this approach. Sometimes, researchers justify their boundary choices in writing. But even then, many innovation systems are bounded by convenience. Indeed, this has been so common that Edquist and Björn Johnson (1997) felt the need clarify the theory involved. They argued that the boundaries around an innovation system are 'always' (Edquist and Johnson, 1997, p 60) or at least 'normally' (1997, p 40) defined in terms of institutions. In other words, boundary specification is meant to be guided by institution theory.

This often does not happen because there are two different uses of the term 'institution' in the innovation studies literature (Edquist, 2001; Coriat and Weinstein, 2002; Grønning, 2008). Some work uses the term for a special category of organizations (Edquist and Johnson, 1997; Grønning, 2008). Here, 'institution' is a euphemism for various public organizations, especially public research organizations (Coriat and Weinstein, 2002). This is linguistically accurate. But at this point it should be clear why I would oppose that use of the term 'institution'. It lumps a variety of public organizations into the support function of an innovation system. It helps us assume an *a priori* role for certain types of organization. Unfortunately, this is the approach that has been taken by prominent innovation scholars like Freeman, Nelson, and Rosenberg (Edquist, 1997). But some scholars have argued for a more theory-laden use of the word 'institution'. Lundvall, Edquist, and others have engaged directly with institutional economics, focusing on the formal and informal rules and routines that shape organizational interaction (Edquist and Johnson, 1997).

Those who have tried to untangle this issue (see Edquist and Johnson, 1997; Coriat and Weinstein, 2002) have turned to Nobel Laureate Douglass North (1990). North argued for a distinction between manifest institutions (that is, organizations) and abstract institutions (that is, rules/routines). Invoking a sports metaphor, he said that 'what must be clearly differentiated are the rules from the players' (North, 1990, p 4). North argued that because any institutional theory must 'begin with the individual' (1990, p 5) and focus on 'groups of individuals bound by some common purpose to achieve objectives' (1990, p 5). In short, his approach encourages the separation of organizations, professions, and other social groupings from the explicit and implicit rules they follow. Edquist and Johnson (Edquist, 1997, 2004; Edquist and Johnson, 1997) argued that research on innovation systems should be grounded in North's institutional theory approach. Based on North's work, Edquist and Johnson (1997) concluded that we should 'deduct' legally constituted organizations from the interactive learning that occurs outside or between organizational entities. Similarly, Casper et al discuss a separation between institutions and 'the interaction of individuals and groups within a particular institutional setting' (2005, p 197). If we fully deploy North's (1990) sports metaphor to an innovation system, we can think of organizations as the players, institutions as the rules, and interactive learning as the sport or the play. Since the work of Edquist and Johnson (1997), there has been an 'increasing consensus' within the innovation systems literature that institutions and organizations should be defined in this way (Grønning, 2008).

So now we have organizational players taking to the field. What sport do they play? Lundvall (1992) describes the innovation systems approach as a 'focusing device' – a kind of social scientific theory – that places our attention on different processes of 'interactive learning'. This emphasis arises from an underlying assumption that 'the most fundamental resource in the modern economy is knowledge, and, accordingly, that the most important process is learning' (Lundvall, 1992, p 1). Note that Lundvall's 'focusing device' does not point towards the noun *knowledge* or any other static outcome. Instead, Lundvall's (1988, 1992) earliest contribution to the innovation systems approach was to embed an understanding of innovation as a verb: an ongoing, ubiquitous, and cumulative learning process. Meeus and Oerlemans explain that, in the context of the literature on innovation systems, learning is 'a process in which all kinds of knowledge are (re-)combined to form something new' (2005, p 159). And so, the contents of innovation systems have been increasingly conceived as networks. As we will see later, these are not networks in the rhizomatic poststructural sense. 'Networks' in innovation studies are almost exclusively plotted within Euclidean, regional spaces. This is positivist, realist, modernist thinking. Social network analysis is the primary analytical tool. It is used to chart the learning relations within and between geographical regions.

In summary, an innovation system can be described as an institutional field in which interactive learning takes place. The boundaries of that field typically correspond to national borders (Meeus and Oerlemans, 2005), regional economies (Gertler, 2010), or sociotechnical regimes (Fuenfschilling and Truffer, 2014). Whether these boundaries are geopolitical, socioeconomic, or sociotechnical, the institutional rules within the system – the cultural norms or laws and regulations (North, 1990) – must be relatively homogeneous. This relative homogeneity can be justified even if we engage in richer institution theory and think about institutions as isomorphic pressures (DiMaggio and Powell, 1983; Irwin et al, 2021). Now, this is not to deny 'institutional work' (Lawrence and Suddaby, 2006). For example, Stine Grodal (2017) has shown how the boundaries around the field of nanotechnology were constructed by various communities over time. However, a relatively static view of institutions has proven useful for innovation systems research. Institutions become part of the 'context' for a case study. Holding them constant lets us move institutions into the background and interactive learning into the foreground. Furthermore, it lets us compare regional contexts. It is the relative homogeneity within one system that sets it apart from the next. This is the essence of the approach and it is not without value. Innovation systems are zones on a map, full of rules, organizations, and interactions. Their boundaries are the points at which the rules change.

But the deeper we dig into institutional theory, the more we start to see that innovation systems must intersect one another (Freeman, 2002; Castellacci, 2009). It is common to acknowledge that institutional fields are nested and overlapping. And this is abundantly clear in research on any 'regional, sectoral innovation system' such as a regional biotech industry (for example, Cooke, 2002) (or a regional ocean tech industry). Some studies, like that by Belussi et al (2010), will fix the boundaries of an innovation system and then explore the openness of those boundaries. Yet, this problem of regional boundary specification has been one of the major unresolved issues in innovation research (Doloreux and Parto, 2005). David Doloreux and Saeed Parto (2005) called it the 'unit of analysis' problem. It is the problem of determining whether innovation system boundaries align with a city, metropolitan region, local district, subnational region, and so on. They proposed that proper deployment of institutions – the 'key variable' in regional innovation systems (Doloreux and Parto, 2005) – helps resolve this issue. But they also warned that 'there is a danger of getting "lost in the woods" while searching for the institutional component' (Doloreux and Parto, 2005, p 146). And this is where the boundaries around innovation systems start to crumble.

Regions matter (Chaminade and Plechero, 2015), but innovation system boundaries can only ever be semi-coherent (Fuenfschilling and Truffer, 2014). We have already seen Jerry Davis' (2022, p 44) argument that the biggest technology firms today are 'indifferent' to sectoral boundaries. Now, consider John Allen's decade-old, matter-of-fact observation that 'parts of global cities like New York and London, for instance, predominantly the corporate finance sectors, are seen to be partially detached from the geographically circumscribed authority of the state' (Allen, 2011b, p 287). Truly, it is hard to imagine any innovation system boundary that is closed off from the outside world (Edquist, 2001). This is because institutions and learning interactions are not necessarily bounded within one geopolitical or sociotechnical space. Innovation processes need not conform with the regions drawn on any map, or with the sectors described in any taxonomy. No amount of institution theory will help innovation researchers resolve this dilemma. One simply cannot have a 'system' without an inside and an outside. To study an 'innovation system', one must establish an analytical boundary. In short, innovation researchers are preoccupied with finding the edges of their subject matter, but have a sense that these edges are more of an analytical tool than an empirical reality. The limitation is paradigmatic.

Testing regional boundaries

What does this mean in practice? Let's play within the modernist paradigm a bit longer and consider my Canadian context. Where many European innovation scholars default to 'national' boundaries, and must then grapple with multinational European Union policy, there is a different subnational challenge to innovation system boundaries in Canada (Holbrook and Wolfe, 2000). This country is a federation of dramatically different socioconomic regions. Under the Canadian Constitution, the individual Provinces retained regulatory power over nearly all industries, businesses, and professions. If you are not from here, then you cannot navigate this dilemma without a map (refer back to Figure 1) (but maps aren't everything). In 'Atlantic' Canada alone, one must consider whether the three 'maritime' provinces are one region or many (Holbrook and Wolfe, 2000) – and whether the province of Newfoundland and Labrador should count at all (it is physically and socioculturally distant). There is also increasing emphasis on major cities as the appropriate contexts for regional innovation system research in Canada (Wolfe, 2014).

I noted this dilemma in the introduction to this chapter. I could have chosen to study ocean science and technology in the City of Halifax or the City of St John's. I had to choose between one province or many. Some readers might set all this aside and argue that the appropriate boundaries for any study of ocean science and technology should correspond with the topography of the ocean. For example, one might argue for a Northwest Atlantic region. And I *could* argue that the ocean has material agency. However, I would quickly hear back from some mainstream reviewer that the ocean does not impose institutional rules in the way that provincial or national governments do.

The stop-gap solution in my PhD thesis was to offload the dilemma on others. I chose to study ocean science instrumentalities more or less around the region where I live. Then I devised a way to avoid any debate over where the exact boundaries might be found. Based on media reports, attendance at sector events, and referrals through personal networks, I identified all the individuals based in Halifax whose work responsibilities included understanding and supporting the ocean science and technology sector. I called these five individuals my 'system experts'. Three of them held industry policy roles – one in a federal government organization and two in different provincial agencies. The other two individuals held ocean technology sector support roles in separate not-for-profit organizations. All five of these 'system-level key informants' (Borgatti et al, 2013) agreed to participate in an interview. During these highly structured interviews, the respondents and I logged all their responses on a touchscreen laptop running the network data collection platform EgoWeb 2.0 (Kennedy and McCarty, 2016). This was all performed in a way that would be acceptable within a positivist social science paradigm.

I asked these 'experts' to identify organizations involved in using and producing ocean science instrumentalities 'in this region'. I deliberately avoiding defining what I meant by 'region'. Once they listed the 'players', I was able to infer where each of them thought the playing field might be located.

I did define the sociotechnical field: ocean science instrumentalities. Onscreen and aloud, I explained that my focus included scientific instruments (such as hydrophones that can be used for collecting data on marine life) and research techniques (such as methods for processing data from those hydrophones) but not new marketing or organizational techniques (such as the way hydrophones are packaged for sale, or the way human resources are managed). Even here, I intentionally left vague boundaries around 'ocean science'. As we will see at the end of this chapter, the field of 'ocean science' is no easier to pin down than the socioeconomic 'region'.

Thankfully, the experts had no problem naming organizations that use and produce ocean science instrumentalities. I engaged them in a free-recall exercise where they could name as many organizations as they wished. They each worked independently to name the relevant organizations, except for one expert who engaged an outside colleague in helping to establish what they felt was a complete list. Following the best practices in network data collection, I used elicitation prompts to ensure that different kinds of organizations were identified (Brewer, 2000). The experts were prompted to include academic, private-sector, government, and not-for-profit organizations. I also anticipated issues in defining 'organization'. So, I explained the concept of a 'kind-of-activity unit':

> For the purposes of this study, an organization is not necessarily a standalone legal entity. In many cases, the parent organization (e.g., Saint Mary's University) is less relevant to this study than a particular department, unit, or division (e.g., the Sobey School of Business). An operating unit can be considered an 'organization' if it engages in one kind of activity and has some decision-making autonomy. (OECD, 2005)

This helped sort out some issues with different units of Dalhousie University, the Nova Scotia Community College, and a few multinational enterprises. Notice how this sidestepped many of the 'organism' metaphor issues raised in Chapter 5.

By the end, the regional boundary was clear. All five experts named a key organization outside the city of Halifax. Four experts restricted their lists to organizations in Nova Scotia. The fifth expert included organizations located elsewhere in Atlantic Canada, including organizations in New Brunswick and Newfoundland and Labrador. Following the majority opinion of these experts, I decided that the innovation system's regional boundary must be the Province of Nova Scotia (it didn't hurt that my gut was already telling me this or that this narrower boundary would save on travel costs).

I also followed the majority opinion in defining the list of 'players' within the innovation system. In total, the five experts named 126 organizations: 60 public organizations and 66 private companies. All the academic, government, and not-for-profit organizations that were named by the experts were 'public' organizations, based on the definitional criteria set out by Perry and Rainey (1988). All five experts independently named the same 11 organizations. An additional six organizations were named by four experts, and ten more were named by three experts. The first four experts named between 25 and 40 organizations each, while the fifth expert named 92 organizations. The fifth expert named disproportionately more organizations because they defined the regional boundary much wider than the other four experts. All these details are represented in Tables 1 and 2. Again, following the majority opinion, I created a 'fixed list' of organizations that comprised the 'core' of my innovation system. All those organizations that were named by three or more experts were included in the list.

Mapping the system

I identified key informants for all 27 organizations on the fixed list. The key informants were either the head of the organization (such as the President, CEO, or Executive Director) or a vice-head with sufficient knowledge of the organization's research and development activities (such as a Vice-President or Director). These individuals were invited to participate in an interview and presented with a research consent agreement based on the best practices

Table 1: Levels of agreement among experts

Level of expert agreement	Organizations
No agreement (one expert: 20%)	72
Two experts agree (40% agreement)	27
Three experts agree (60%)	10
Four experts agree (80%)	6
All five experts agree (100%)	11
Total organizations named	126

Table 2: Number of ocean science instrumentality organizations identified by experts

Expert	No. of organizations	Agreement with other experts		
		Two or more others	One other	No others
#1	32	18	6	8
#2	25	19	5	1
#3	28	17	7	4
#4	40	19	15	6
#5	92	23	16	53
Total	126	27	27	72

Note: Totals indicate the number of unique organizations named.

template developed by Borgatti and Molina (2005) and approved by the research ethics boards at both Saint Mary's University and Acadia University. With respect to confidentiality, the consent agreement specified:

All data obtained from private sector companies will be kept confidential and will only be reported in an aggregate format (by reporting only combined statistics and by representing all private companies using one common colour/shape on network diagrams). No one other than the primary investigator and supervisor listed above will have access to the data about individual interviewees and the data about private companies. Data about public and not-for-profit organizations will be treated as public-record (i.e., not confidential), except where relationships with private sector companies are noted. To protect the strategic interests of private companies, this data will remain confidential.

The consent agreement included a sample network graph which was used at the outset of each interview to explain the risk that private sector organizations may be identifiable in the research outputs. Given sufficient knowledge of the research context, an informed reader may be able to infer the names of different companies from their relations or positions on a network graph (Borgatti et al, 2013). This consent agreement was signed by all participants.

I conducted face-to-face interviews with key informants from 25 of the 27 organizations on the fixed list. Only two of the 27 organizations on the fixed list did not participate in an interview. After multiple interview requests, senior officials at one private company and one academic PRO did not respond. To maintain confidentiality, these two organizations are unnamed in my work.

As with the preliminary round of expert interviews, the organizational key informant interviews were conducted using a touchscreen computer running the network data collection platform EgoWeb 2.0 (Kennedy and McCarty, 2016). Draft interview questions were reviewed by the five system-level experts and then revised to clarify language, expedite data entry, and better reflect the context of ocean science and technology. The same structured interview questions were used for all types of organizations – private companies, PROs, other public organizations, and not-for-profit organizations. Interviews ranged from approximately 30–90 minutes in length.

First, I asked respondents simple questions about the type of organization they were representing: academic, government research, other government, private company, or not-for-profit; the total number of full-time equivalent employees working at the organization, and the number working in R&D; the kinds of outputs the organization produced over the previous five years and the novelty of those outputs; and whether there had been any changes to the way these outputs were produced over the previous five years, and the novelty of those changes. These questions were the preamble to the more complex task of network data collection.

Most of the interview followed a standard 'personal-network research design' (Borgatti et al, 2013). This is a process to produce ego-network data: data on the network of alters, or relations, around each ego, or focal organization. A standard personal-network interview instrument includes two phases of questions: a name generator to establish a list of alters, followed by name interpreter questions to collect data about the alters and about ego's relationships with them. For the name generator, I presented each respondent with a roster that included the 27 organizations on the fixed list, plus the 20 additional organizations that were named by only two experts. Respondents were asked to review the roster and identify those organizations that their own organization usually interacted with over the past five years. Then, during the name interpreter, the EgoWeb 2.0 software (Kennedy and McCarty, 2016) produced a grid where all the selected organizations appeared as rows and seven different types of interactive learning relationships appeared as columns (see Figure 2). These seven types of interactive learning relationships were adapted from the work of my PhD supervisor, Claudia De Fuentes (see De Fuentes and Dutrenit, 2012). I only needed to expand her model to add the transfer and sharing of equipment and technical services. All those relationships that could have directionality were presented twice on the screen during my interviews. For example, respondents could say that they licensed or transferred intellectual property *to* an alter organization, and/or that they licensed or transferred intellectual property *from* an alter organization. This meant that respondents could choose from among the ten different interactions listed in Figure 2 and could select all that applied.

Figure 2: The multigrid interactive learning relationships component of the interview instrument

The table below includes all of the organizations you named (in the rows) and various ways that Organization may have interacted with them (in the columns).

Please check as many types of interaction as are relevant for each organization. The next screen will provide an opportunity for you to add any additional types of interaction that I may have missed.

	We had a formal R&D contract, partnership, or sponsorship with...	We licensed or transferred intellectual property to...	We licensed or transferred intellectual property from...	We acquired equipment /services from...	We provided equipment /services to...	We formally shared information with...	We maintained informal relationships with...	Knowledgeable individuals moved here from...	Knowledgeable individuals left here for...	We shared knowledgeable individuals with...
Organization A	☐	☐	☐	☐	☐	☐	☐	☐	☐	☐
Organization B	☐	☐	☐	☐	☐	☐	☐	☐	☐	☐
Organization C	☐	☐	☐	☐	☐	☐	☐	☐	☐	☐
Organization D	☐	☐	☐	☐	☐	☐	☐	☐	☐	☐
Set All	☐									

Back Next

Source: Produced by author using EgoWeb 2.0 (2016).

Personal-network research designs sometimes include a third phase of 'name interrelator' questions. This is where one might collect data on the connections between alters. However, this introduces considerable interviewee burden because it is time-intensive (Borgatti et al, 2013). Rather than using a name interrelator to collect nearly identical whole networks from each respondent, I used the common approach of aggregating ego-alter data across all respondents (Borgatti et al, 2013). This involved overlaying and merging network data from multiple respondents into one whole network that includes all respondents and the relations among them. From a positivist perspective, a convenient by-product of this data fusion process is triangulation of the network data (Borgatti et al, 2013). It also provides an opportunity to reliably reconstruct minimal levels of missing data (Stork and Richards, 1992; Huisman, 2009). This reconstruction is possible because the responses provided by each respondent overlap with missing values from each nonrespondent (Huisman, 2009). Although two of the 27 organizations on my fixed list did not participate in an interview, all 25 participating organizations provided data on their relations, or lack of relations, with the nonrespondent organizations. These two nodes can therefore be included in the network, as data were collected on 25 of their 26 possible relationships. Only two of the 702 possible relational paths – 0.003 per cent of the observations – were missing from the whole network dataset. Rather than computationally estimating the missing observations, I was able to run all analyses twice – turning the missing path 'on' or 'off'. Since the results did not change, I followed the conservative assumption that these two organizations had no interactive learning relationships.

In the end, I produced a network graph of interactive learning in ocean science instrumentalities around Nova Scotia. I mapped one 'strongly-connected component' that comprises 27 organizations (see Figure 3). In other words, no organizations were isolated and all organizations were reachable through paths of interorganizational interactive learning relationships. The network included 12 scientific instrumentality companies and ten PROs. The ten PROs are listed and described in Table 3. The network also included five organizations that were highly engaged in ocean science, but did not directly engage in scientific investigations. One of these is a teaching unit of the Nova Scotia Community College. Four of these are not-for-profit organizations that meet two of the three criteria developed by Perry and Rainey (1988) for being classified as public organizations. I therefore labelled all five of these organizations as public support organizations.

This brings us to what I considered the most important feature of my data collection process: the equal treatment of public and private organizations. The roster of organizations that was identified by system-level experts included any type of organization that fell within the sociotechnical field. Then, key informants with those organizations all responded to the same

Figure 3: Interactive learning network for ocean science instrumentalities in Nova Scotia, Canada

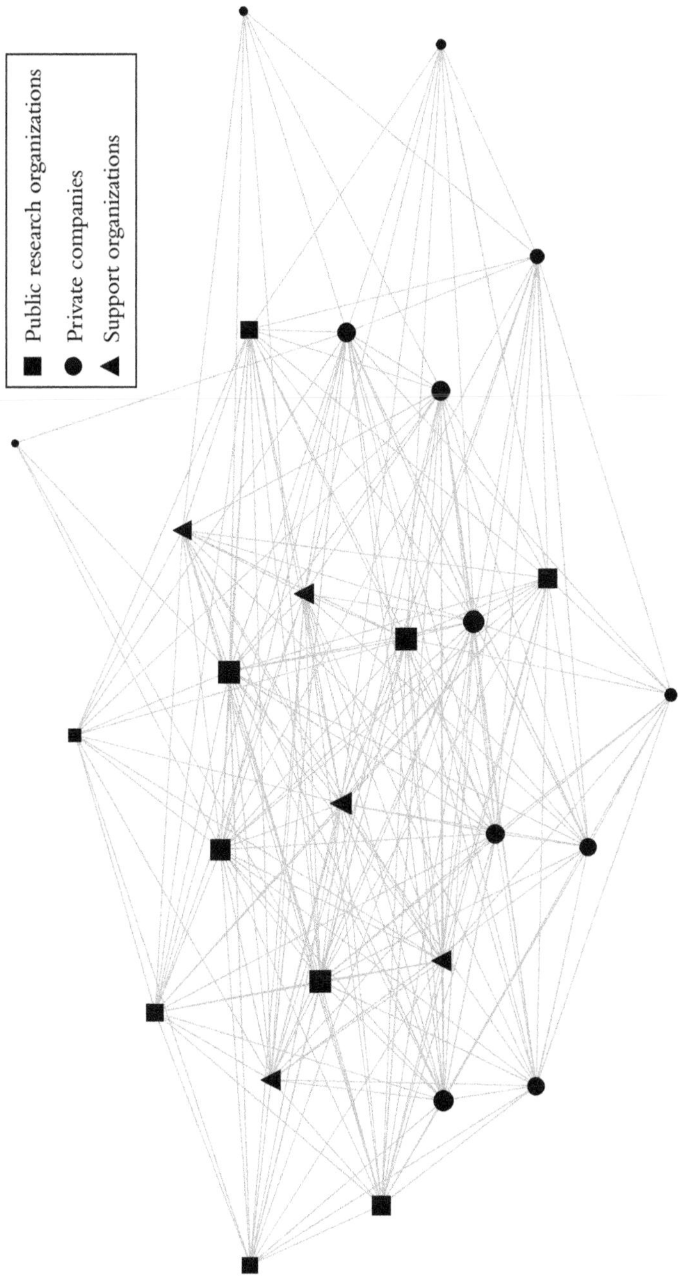

Public research organizations
Private companies
Support organizations

Note: Nodes sized by degree.

Source: Graph produced by author in NetDraw (Borgatti, 2002).

Table 3: Public research organizations in the interactive learning network

Organization	FTEs	R&D Intensity (%)
Acadia Tidal Energy Institute	11	98
Bedford Institute of Oceanography (Department of Fisheries and Oceans)	700	21
Bedford Institute of Oceanography (Natural Resources Canada)	55	82
Verschuren Institute, Cape Breton University	40	90
Oceanography Department, Dalhousie University	118	97
Defense Research and Development Canada: Atlantic Research Centre	165	61
Applied Geomatics Research Group, Nova Scotia Community College	20	75
Applied Oceans Research Group, Nova Scotia Community College	10	100
Ocean Tracking Network, Dalhousie University	12	88
Academic Kind-of-Activity Unit (non-participant)	–	–

Note: FTEs = full-time equivalent employees (a measure of size). All organizations in this table are public, according to the criteria developed by Perry and Rainey (1988): they are all under public ownership, receive public funding, and operate under polyarchal social control. FTEs and R&D intensity for the nonparticipating PRO were available from online sources, but are suppressed in this table to maintain confidentiality.

structured interview, regardless of any public/private sector distinctions. Because all organizations were treated equally for data collection, I could test some hypotheses and produce some statistics about public innovation in goods. I will return to the data that fell within my survey boundaries and how I analysed it in the next chapter. But first, we should consider the hints of data that did not conform with a regional-network approach. This concluding discussion is not a confessional account of my technical errors (which are likely plentiful) or my poor use of method.[1] Rather, I turn to the more fundamental question of how *regional* and *network* spatial metaphors focus our attention on some aspects of innovation, but not others.

Topological alternatives

Thus far in this chapter, I have surveyed the boundaries around an innovation system, laying the groundwork for an innovation survey. I defaulted to the *regional* type of spatial reasoning that underpins the systems of innovation literature. To think about innovation systems, we must first think about space as measurable, divisible, *Euclidean* fields. We saw that, in the mainline innovation literature, institutional fields are conceived as arenas for interactive learning networks/relationships. Whatever the 'size' of the regional space, it is *volumetric*. It contains and captures certain kinds of innovation, in defined fields of technology, and in particular places. Boundaries can be plotted. In my case, this spatial reasoning enabled a survey of organizational learning interactions, involving the production and use of ocean science instrumentalities, in Nova Scotia, Canada.

The regional/volumetric approach is adopted widely and without question in innovation studies. Any debate is about scale; it is about the size of the regional volume(s) under investigation. And we have seen that the debate about scale is really a search for the 'institutional component' of an innovation system (Doloreux and Parto, 2005, p 146). This is where innovation researchers are getting 'lost in the woods' (Doloreux and Parto, 2005, p 146). I have suggested that innovation research could use more sophisticated institutional theory tools, and these might provide a less static and bounded perspective. But this will only make it more difficult to trace the 'edges' of an innovation system. An alternative, which we began exploring in Chapter 5, would be to change our metaphors and thereby engage in different puzzle-solving activities. And I have already argued, in Chapter 3, that context is not a container. So perhaps innovation research could learn from a more radical set of spatial metaphors? Let us consider new ontological instruments – intellectual tools borrowed from social topology – where regions and networks are only two of the many options for thinking about space.

Topology is the study of how we understand and represent spatial relations. It is conceptually borrowed from mathematics, where methods were devised to account for geometric objects that can be stretched, folded, or otherwise deformed. It is silly to mathematically pin down the corners of a rubber sheet if you are interested in how its surface bends and warps. Topological shifts have therefore proven useful for understanding mathematical problems where the solutions are obscured by a focus on the absolute position and size of objects in space (and time). For example, physicists are making frantic use of alternative topological reasoning in their research on dark matter (for example, Derevianko and Pospelov, 2014; Afach et al, 2021). This was the subject of the 2016 Nobel Prize in Physics. Many believe that dark matter and dark energy might only be observable through topological defects: wrinkles in gravity or time. Less abstractly, topology was useful in the 1700s, when Leonard Euler shifted the famous Königsberg bridges problem from regional to network space. By disregarding the absolute position of the seven bridges around the city, Euler produced a proof showing that it was impossible to plot a route that crosses each bridge only once (Barabási, 2013). This might seem like a trivial maths puzzle, but Euler's spatial reasoning is now extremely useful in everyday life. Consider the topological map of the

London Underground. It shifts attention away from the confusing twists, turns, rises, and dips of the underground tunnels and instead focuses riders on the routes and connections between stations. This is an easier way to understand how the subway system moves beneath the city. It is a map of network space. It doesn't deny that the trains move through real physical space; rather, it demonstrates that there are other useful ways of understanding our sociomaterial world.

This is why topology has become 'one of the putative core topics in geography' (Lata and Minca, 2016, p 439). As John Allen observes, 'something seems to be happening to the way that we think about space and time – as non-linear, intensive, folded even – that increasingly chimes with our experience of the world' (Allen, 2011a, p 317). In his work, topology helps reveal the non-Euclidean ways in which nongovernmental organizations (NGOs), governmental organizations, and multinational corporations exert power remotely and over great distances. Beyond questions of geopolitical power, Allen has argued that topology provides poststructuralist geographers with a 'looser, less rigid approach to space and time that allows for events elsewhere to be folded into the here and now of daily life' (Allen, 2011b, p 283). This has enlivened thinking about 'local' and 'global' relations (Lata and Minca, 2016; Latham, 2011). Indeed, Alan Latham has argued that 'topological notions of space-time are most useful when they are used to challenge the very idea that there is a "global"' (2011, p 315). In other words, topology can help us problematize the notion of 'scale' (Asdal, 2020; Oppenheim, 2020). It helps us tackle 'the elusive character of borders, scales, territories, regions or networks' (Lata and Minca, 2016, p 440). This is why the geographer of boundaries Anssi Paasi (2011) has argued for 'a need to move from "absolute" to relative and relational space or from Euclidean metric spaces to some *other* spaces' (Paasi, 2011, p 300, emphasis added). Notice here that the word 'absolute' stands in for a modern, realist ontology. Paasi's (2011) alternatives are relative/postmodern and relational/amodern.

When geographers have imagined these 'other' spaces, many have turned to the rhizomatic plateaus of Deleuze and Guattari (1987) and to the material-semiotics of ANT (Lata and Minca, 2016). The 'topological ethos' of John Law and Annemarie Mol (2001) has been an especially 'prolific' source for new theorizing in geography (Lata and Minca, 2016). Organization studies has also benefited from very similar 'object lessons' offered by Law and Singleton (2005). (The same arguments were presented a decade earlier in STS: see Mol and Law, 1994.) In these various contributions (Mol and Law, 1994; Law and Mol, 2001; Law and Singleton, 2005), Law, Mol, and Singleton argue that our default topologies suggest stability and spatial integrity – impeding other ways to make sense of sociomaterial dynamics. Those default topological metaphors are the region and the network. Law, Mol, and Singleton advocate for 'fluid' and 'fire' as alternate metaphors.

A brief discussion of these four topological metaphors will help conclude my argument.

Regions

First, we default to thinking in regions. Research objects appear stable and grounded (pardon the pun) when they are framed by a regional topology. This is the 'common sense' view where 'we tend to think of objects as physically constituted items that occupy a volume in Euclidian space' (Law and Singleton, 2005, p 335). And yet, 'regionalism' is undoubtedly a set of 'topological rules about areal integrity and change' (Law, 1999, p 6). The rules of regionalism are socially constructed and contested. Much of this chapter has been a description of the rules that I applied to set boundaries around a regional innovation system. In making the rules explicit, we can see how 'objects are clustered together and boundaries are drawn around each cluster' (Mol and Law, 1994, p 643). We can also see past the rules to the underlying assumption: we are working with a surface or volume that must be 'broken up into principalities of varying sizes' (Law, 1999, p 6). Of course, this topological assumption extends well beyond regions on a map – it helps us divide all manner of sociomaterial objects.

This topology is especially suited to discussions of national and regional policy, provided that the principality of the state aligns with the principality of the policy phenomena. But here is where innovation research must depart from regional thinking. Innovation is not geographically stable. We might sometimes think about how innovation is anchored – around a large R&D organization (Niosi and Zhegu, 2010) or a key piece of instrumentation like the Aquatron in Halifax, or the geotechnical centrifuge in St John's that I mentioned at the beginning of this chapter. Innovation can be understood regionally, but we know it is not regionally delimited.

Networks

The network metaphor helps us think without regional limits. I have also approached the idea of a network topology in this chapter, but not in the way that is common to STS. In STS, and then other fields, ANT provided a topological understanding where 'elements retain their spatial integrity *by virtue of their position in a set of links or relations*' (Law, 1999, p 8, emphasis in original). In other words, the integrity of an object is maintained through a pattern of sociomaterial links that remain stable across space and time. Bits and pieces of scientific knowledge – instruments, diagrams, texts, etc. – can carry action across distances if they become punctuated as a 'black box' (Latour, 1987). They can be thought of as 'immutable mobiles' (Latour, 1987) – objects that transcend physical space without losing their shape.

This is the network/rhizome topology – an ontological instrumentality – provided by ANT.

I noticed many immutable mobiles of ocean science during my data collection. At the Applied Oceans Research Group in the Nova Scotia Community College, there was a big round device called a CTD rosette (for conductivity, temperature, and depth), loaded with a variety of sensors and sampling apparatus. One would need at least a pick-up truck to transport this device, which is lowered into the ocean to collect samples and data at various predetermined depths. It was that lab's turn to physically house the device – but its logoed surface was like the back bumper of some American cars. It was loaded with stickers from ocean science and technology organizations nearby and afar. This was a well-travelled device. I also saw and heard about many devices whose relations stretched beyond Nova Scotia. Some of the science organizations I surveyed had important instrumentation partners outside my research boundaries. Similarly, some of the instrumentation companies I surveyed had their principal scientific partners elsewhere in the world. However, none of this became data. None of these relations found their way into my study and none will appear in Chapter 7 of this book. When I had established *my* boundary – defined *my* innovation system – I had imposed a Procrustean transformation on the relations of ocean science technologies. I established the size of the 'bed' and then crafted a survey instrument that would deftly, rigorously, but quietly cut off any rhizomatic shoots. I captured network data, yet I did so within a regional/volumetric topology.

The network topology common to STS – the one that is rhizomatic, not regional – is suited to understanding how a scientific instrumentality is taken up and used from one laboratory to the next. Indeed, that network topology has its 'roots' in the ANT laboratory studies. It helps us think about the translation of objects through sociomaterial space and time. It was useful earlier in this book when I explored different ways of knowing the past. But notice that this topology is not about plotting relations between organizations or laboratories – as I have done in this chapter and as is common in 'network' analyses of innovation. That modernist network approach is about understanding the geometry of relations contained within a region. It is a form of regionalism. An amodern ANT-inspired network topology is 'not about a volume within a larger Euclidean volume' (Law, 1999, p 6). In this way the rhizomatic ideals of ANT actually 'helped destabilize Euclidianism' in many fields of social science (Law, 1999, p 8).

And so, we know that technology and innovation can be understood differently as network phenomena. There is a tremendous literature on this in STS and I agree with Martin (2016) that innovation studies must catch up with that work. Yet, many STS scholars have moved on. By the late 1990s, it had become clear that the network metaphor 'had the effect of limiting the conditions of spatial and relational possibility' (Law, 1999, p 8). More recently, in the journal *Organization*, Tommy Jensen and Johan Sandström have argued that 'ANT is still haunted by a conceptual conundrum: while opening up for spatial complexity, the lure of the *Network* risks drawing representations into a singular, network space, "othering" spaces that might be critical for organizing' (2020, p 702, emphasis in original). In that same journal 15 years earlier, Law and Singleton (2005) had characterized the ANT network topology as 'an inappropriately rigid and centred version of relations' (2005, p 341). This is where they advocated for fluid and fire as alternate topological metaphors.

Fluids

Regions appear stable because we position objects within Euclidean space-time. Networks appear stable because we position objects in relation to one another. But the idea of 'fluid' space dispenses with stability altogether.

The 'fluid' metaphor was developed in a study of 'the Zimbabwe Bush Pump' by Marianne de Laet and Annemarie Mol (2000). Many people speak of this pump as a fixed entity, but de Laet and Mol (2000) describe how instances of the pump are different from one another in interesting and incremental ways. Key to the success of the pump is the way in which the inventor and manufacturer 'dissolve' their 'actorship' or 'authorship' (de Laet and Mol, 2000, p 249). No one controls the various sociomaterialities we might call a bush pump. No one polices the boundaries of how a pump should be installed, used, and repaired. This is why de Laet and Mol say that 'the Zimbabwe Bush Pump is solid and mechanical and yet ... its *boundaries* are vague and moving, rather than being clear or fixed' (de Laet and Mol, 2000, p 225, emphasis in original). Law and Mol (2001) later explained that, in fluid space, multiple instances of an object are the same, but not identical. Because one object flows into the next, 'a fluid world is a world of *mixtures*. Mixtures that can sometimes be separated. But not always, not necessarily' (Mol and Law, 1994, p 660, emphasis in original). This means that 'in fluid spaces there are often, perhaps usually, no clear *boundaries*' (Mol and Law, 1994, p 659, emphasis in original). For some readers, this topology might seem unreal. But consider how *unreal* it is to split and label our planet's one ocean into sections. Andrea Ballestero (2019) has produced a wonderful book on some of the devices we use to separate water: formulas, indices, lists, and pacts. A fluid topology helps us see how water flows through us and this planet.

In practice, a fluid topology can reveal some of what I missed by surveying stable boundaries. For example, I needed to stabilize 'ocean science' so I could use it as an innovation system boundary. But we could see multiple, interrelated understandings of ocean science and ocean engineering from the very beginning of this chapter (and also Chapters 3 and 4). If we were to switch into other languages, the differences would be more self-evident. The Russian term translates to oceanology rather than oceanography and thereby denotes important differences in the scientific practices (Hamblin, 2005; Mills, 2011). Despite those apparent differences, we saw in Chapter 4 that Russian oceanology was not separate from Canadian oceanography – even during the Cold War. In his autobiographical history of oceanography, William Bascom notes that 'oceanography is not so much a science as a collection of scientists' (1988, p xiii). Depending on where you sit, ocean science might be focused on physical and/or biological and/or chemical processes. Benson and Rehbock go further, saying, 'oceanography is a hybrid, a mixed science ... [that] cannot be said to be a single scientific discipline' (1993, p ix). Fluidity is evident in these scientific practices.

At issue is the multiplicity with which humans relate to the ocean. For example, Bascom credits four factors with the rapid growth of ocean science after the Second World War: a 'doubling' in submarine warfare, a 'tripling' of the global fish catch, the shift to *offshore* oil production, and a new public interest in marine conservation and archaeology (Bascom, 1988, p xiv). Eric Mills' history of the field uses a slightly longer list. He writes that due to demand from 'fisheries, shipping, sewage disposal, ocean mineral exploitation, and submarine warfare, the field [of ocean science] had expanded too rapidly for the supply of personnel from the pure sciences to keep pace' (Mills, 2011, p 254). Ocean science is enacted in so many ways that disciplinary boundaries are problematic. We saw this in Chapter 4, with the challenges Dalhousie University faced advocating a biological focus against the University of British Columbia's physical oceanography focus. Eventually, Dalhousie found some value in a 'fluid' collection of ocean scientists. But even then, the ocean does not provide a hard disciplinary boundary. Notice that ocean science and technology do not always stop at the shoreline: some ocean science and technology travel into space, and vice versa. If I had been situated in Massachusetts, like Helmreich (2009), I would have encountered the astrobiologists who are mixed into the Woods Hole Oceanographic Institution. The fluid topological metaphor helps us notice this endless mixing.

Like Mol and Law (1994), I must try not to favour the fluid metaphor too much here. They say that 'fluid spaces are no "better" than regions or networks' (Mol and Law, 1994, p 663). They simply give us a different understanding – an alternative to the Procrustean transformations of regionalism and relational rigidities of networks. Some say that a fluid topology gives us gradual and incremental change (Law and Mol, 2001). And we sometimes write and talk about innovation in this way. But I favour the way in which Law unpacks the metaphor differently in *After Method*:

> in this way of thinking the world is not a structure, something we can map with our social science charts. We might think of it, instead, as a

maelstrom or a tide-rip. Imagine that it is filled with currents, eddies, flows, vortices, unpredictable changes, storms, and with moments of lull and calm. (Law, 2004, p 7)

This tells me that a fluid topology could be useful for thinking about more than incremental innovation. It could also be applied to breakthrough innovation – after all, water can cause tremendous, sometimes destructive change. But what might we make of the destroyed and the absent? Those things are dissolved, or washed away, with the fluid metaphor. And so, the topology of 'fire' might also have value.

Fire

The topology of 'fire' emphasizes discontinuity (Law and Mol, 2001; Law and Singleton, 2005); it is about instability (vs. region/network) and disconnection (vs. network/fluid). Fire objects are 'energetic and transformative, and depend on difference – for instance between (absent) fuel or cinders and (present) flame' (Law and Singleton, 2005, p 344). Thinking with this metaphor keeps our attention on what must be made absent to make something else present. This becomes a valuable tool when we accept that 'not everything can be brought to presence. Or, to put it differently, to make things present is necessarily also, and at the same time, to make them absent. Presence, in short, depends upon absence (just as absence depends on presence). This is a matter of logic, of definition' (Law and Singleton, 2005, p 341).

Through a 'fire' topology, we might begin to understand the presences and absences that were created during my boundary survey. My attention was on bringing public organizations and scientific instrumentalities into presence. I knew that a focus on the market – as prescribed by the OECD's innovation survey manual (OECD, 2005) – is a problem because it conceals user innovations (Gault, 2012, 2020). I followed the advice of Fred Gault (2012, 2018, 2020) and wrote a survey that would count any innovation that had been brought 'into use' anywhere in society. This was an absence in other research that I made present in my own work.

Of course, we cannot make everything present. Some absences are intentional. For example, I knew that some organizations I surveyed develop secret military technologies that respondents could not legally discuss. But there are also absences that we cannot imagine: the flame is so bright and exciting that it casts a shadow over the 'other'. Some critical scholars of innovation have already argued that the brightness of the pro-innovation bias shifts attention from moments where 'no' or 'slow' innovation might be appropriate – like opportunities for degrowth (for example, Cañibano et al, 2017; Leitner, 2017). Also, consider Vinsel and Russell's (2020) argument that 'innovation' deludes us into ignoring the importance of maintenance and repair. It is clear, then, that the idea of 'innovation' is focused on the 'flame' of novelty and thereby casts shadows over other important socioeconomic phenomena.

After several of my interviews, I noticed that the word 'innovation' was creating these kinds of 'other' interesting absences from my survey data. For example, I was fascinated by a set of oceanographic instruments that were being constructed in a lab at my university. I saw bits and pieces of everyday construction materials, like one might see in the shop of a plumber or carpenter. In this lab, an interdisciplinary team predicts and studies the ecological impacts of tidal energy turbines. Physical oceanographer Brian Sanderson builds and repairs various 'low-cost' drifters for their collaborative studies in the Minas Passage, the Minas Basin, and the Bay of Fundy. In their publications, the research team describes limitations of the more expensive and typical approach: mooring instruments in place (Adams et al, 2019; Sanderson et al, 2021). They are working with the world's highest recorded tides and the tidal flow produces so much flow noise that standard instrument assemblies produce poor data. Sanderson et al drastically reduced flow noise in their data by allowing their instruments to drift with the current and not be disrupted by waves – figuring out how to control that movement with assemblages of plywood (called drogues) that were also carefully fashioned to avoid entangling lobster fishing gear. For one set of studies, instrument drifters were built from '38 mm diameter ABS pipe and common plumbing fittings with flotation fashioned from 50 mm foam board insulation' (Sanderson et al, 2021, p 58). Inside those pipes were batteries, inexpensive GPS trackers designed and marketed for dog owners, and a Nova Scotia-produced Vemco brand acoustic receiver. This receiver is designed to detect 'pings' from Vemco tracking tags which other researchers have surgically implanted into fish, such as Atlantic salmon. The receivers are normally fixed in place (O'Dor et al, 1998), like at the mouth of a river, and are therefore subject to the 'noise' of rushing water and jiggling waves. This inexpensive drifter solution, bootstrapped by scientists in their own lab, took measurements differently and thereby proved far superior to moored devices for certain scientific issues. They have done similarly for hydrophone detection of marine mammals (Adams et al, 2019). Their approach to constructing drifters has even allowed local school children to build their own instrument assemblies and collect high quality scientific data (Redden, 2016). In short, these researchers built ingeniously simple and inexpensive devices so that they could produce novel science. Their science and their instruments are uniquely crafted for a particular purpose at a particular place and time. No piece of their equipment is really 'new to the world' or eligible for patent protection. I would like to call it 'innovation' and yet it would not meet any of the standard OECD survey criteria. So, while their lab appears as a public research organization in my study, there was no space in my innovation survey for the details in this paragraph.

Because I was studying innovation, I could not study non-innovations – the things Godin and Vinck call 'novation' (2017b, p 3). And so, there are always absences like 'novation' that we overlook because our instrumentalities point us another way. Certain forms of novelty are valued above all else. I will turn to one of these absences in my final chapter. But for now, my point has been made. I am suggesting that there is unexplored opportunity to theorize innovation through a fire metaphor. This is not only because 'destruction' and 'discontinuity' are common ways of thinking about innovation; it is also because innovation studies should involve some examination of unrealized potentials. Some of today's absences will become tomorrow's innovations. So, Law and Singleton do not go far enough when they celebrate fire objects for 'their novelty, their creativity, their destructiveness' (2005, p 349). Fire topology also allows us to acknowledge absences – the seemingly non-innovations that are cast into the shadows and the combustive materials that are not yet aflame. This is likely difficult research because fire objects cannot 'be domesticated' (Law and Singleton, 2005, p 347).

Other topological metaphors

In this chapter, I have argued for innovation research to break from its singular topological perspective. The field's most common instrumentalities – collected under the innovation systems approach – rely on regional or volumetric thinking. If you prefer, this has also been called 'arborescent' (Deleuze and Guattari, 1987) thinking. Other topological metaphors can stretch, twist, and distort our observations in other fruitful ways. In the words of Iulian Barba Lata and Claudio Minca, topology gives us 'a spatial lexicon' (2016, p 441) that can account for the intersecting multiplicity of space-time. Surveying multiple topologies will give us multiple understandings.

The four topological metaphors considered here are not the only possibilities. We need others. Some say these ones have serious limits. For example, Allen (2011b) has argued that metaphors like fluid and fire 'serve only to confuse rather than enlighten' (2011b, p 283). He suggests that they are 'failed metaphors, words which, after their initial promise faded, nobody was much interested in using' (Allen, 2011a, p 317). He argues that playing with topological metaphors 'may be colourful, but owes little to the eye-opening possibilities that topology offers' (Allen, 2011a, p 318). He writes well, but seems to miss the point. Working with multiple metaphors means 'we can avoid naturalizing a single spatial form, a single topology' (Law, 1999, p 7). The scholarly task is to expand our 'spatial imagination' through 'metaphorical proliferation' (Latham, 2011, p 315). Thinking about space and place through multiple metaphors provides an ' "intertopological" effect

... a spatiotemporality not only of the connected and unconnected, but also of the potentially, or not yet, connected' (Oppenheim, 2020, p 318).

One approach might be through meta-metaphor. 'Hyperobject' has recently been proposed as a way to appreciate the many topological possibilities for innovation research (Rehn and Örtenblad, 2023). As a hyperobject, innovation can be understood to be 'massively distributed in time and space, to the point where most things can be seen as innovation depending on which spaciotemporal position you choose to occupy' (Rehn and Örtenblad, 2023, pp 6–7). Another approach might be to create space for new understandings through the 'interruption of topology' (O'Doherty, 2013, p 211). Damian O'Doherty (2013) did this in an ethnographic methodological experiment where 'an arbitrary set of rules and constraints following the rigor of a mathematical series of calculations and measurements were devised to generate a sequence of random walks traversing the city of Manchester' (O'Doherty, 2013, p 215). Based on the resulting insights, O'Doherty would have us proceed without any single metaphor: 'bereft of any abstract principle' and therefore open to 'the conditions of possibility for thinking topology' (O'Doherty, 2013, p 226).

The alternative to all this topological play is accepting the 'hegemony' (Sepp, 2012, p 47) of regionalism. That is the innovation studies norm. But in geography, it is normal to investigate – intertopologically – the construction of boundaries, regions, and territories (for example, Sepp, 2012; Asdal, 2020; Oppenheim, 2020). What if I had done that here? While I was busy pinning down the boundaries of ocean science instrumentality innovation, there was work being done to shift public and government attention from a Nova Scotian 'ocean technology cluster' to an Atlantic Canadian 'ocean supercluster' (see Doloreux and Frigon, 2021). And as I write this book, there has been a further shift in the governmental language: Canada's regional 'innovation superclusters' have been rebranded as 'global innovation clusters' (see Sá, 2022). Note the hyperbole, but also the opportunity to apply unstable topologies.

Some innovation scholars might say that we need the region/volumetric metaphor so we can pin things down, survey them, and quantify them – otherwise they do not count. But topology provides 'a way of understanding space and time when the numbers no longer quite add up to anything significant' (Allen, 2011a, p 316). And as we will see in the next chapter, statistically significant results can become meaningless in the face of staunchly held values and beliefs.

7

Sisyphean Statistics

After collecting the survey data described in Chapter 6, and completing my doctoral thesis, I considered never publishing my statistical analyses of public innovation in ocean science instruments. The results sat in a metaphorical 'file drawer' for years after my PhD defence. But this was not the well-known 'file drawer problem' where science is skewed by the suppression of insignificant results (Rosenthal, 1979). Instead, I sat on these statistics because I had come to hate them. I had enjoyed producing them; the work was challenging and stimulating, not routine or 'banal' (Lippert and Verran, 2018). And I was passionate about the point these numbers make – serving as 'evidence' of public innovation in goods and 'proof' of poor public policy in the place where I live. However, statistical conventions would prove antagonistic towards this passion and politics.

It had seemed that statistical evidence was needed to shift public policy. One of my system experts had told me that the recent government cuts to ocean science, especially the cuts at BIO, would be devastating for Nova Scotia. This expert said the most important contribution I might make would be to show this in numbers. And so, I embarked on the fool's errand of trying to debunk neoliberal dogma with statistics. Following convention, I wrapped the statistics in the trappings of rationality and objectivity. I found that the numbers could only thrive if they appeared apolitical. But I also found that depoliticizing the statistics made them trivial.

Like Helen Verran, in the first iteration of her book *Science and an African Logic*, I found that my numbers work 'failed to deliver a useful critique' (Verran, 2001, p 20). And so, this chapter deconstructs the tools and techniques of statistical analysis in innovation studies. I describe statistical analyses, but my analytic tool is autoethnography (Ellis, 2004; Prasad, 2019). This chapter is an inquiry into my own experiences (*auto-*) navigating the culture of innovation statistics (*-ethno-*). The story (*-graphy*) is Sisyphean. I will suggest that following convention is like being condemned by the statistical gods to push numbers up a hill, hoping to successfully reach the summit, only to realize that the effort was ultimately meaningless.

Most critical scholars would eschew quantitative methods; quantitative analysis is firmly embedded in a positivist worldview. However, several critical management scholars have already demonstrated the utility of ethnostatistics (Gephart, 1988, 1997, 2006) in problematizing the assumed objectivity of statistics (for example, Boje et al, 2004; Smith et al, 2004; Helms Mills et al, 2006). Ethnostatistics is 'the empirical study of how professional scholars construct and use statistics and numerals in scholarly research' (Gephart, 2006, p 417). It has seen only limited use in innovation research. Kilduff and Oh (2006) used ethnostatistics to examine four different and conflicting statistical analyses of the same medical innovation diffusion data. They concluded that the differences in statistical analysis are irreconcilable 'given the radical undecidability of numerical evidence in the absence of context' (Kilduff and Oh, 2006, p 432). Aside from a few papers like this, ethnostatistics is a generally underutilized research method (Gephart, 2006; Helms Mills et al, 2006). When it does get used, it is normally aimed at some group of 'others' who are producing and using statistics. In this way, ethnostatistics tends to reproduce the impartial and detached analytical stance – the 'god trick' (Haraway, 1988) – that it aims to disrupt. Autoethnography (Ellis, 2004; Prasad, 2019) offers an alternative. It helps me to bring my situatedness into an analysis of the 'epistemic culture' (Knorr Cetina, 1999) around innovation statistics. It helps me break from what Adam Saifer and Tina Dacin call 'a rather strict and rationalistic understanding of how people actually experience and engage with data and datafication' (2021, p 624). Through autoethnography, I can address 'the feeling of numbers' (Kennedy and Hill, 2018) and 'the aesthetic, emotional, and discursive aspects' (Saifer and Dacin, 2021, p 623) of statistical work.

Of course, the problems embedded in standardized innovation statistics are already a major concern for the field. Gault (2018, 2020), Godin (2002, 2005), and Perani (2019, 2021) have all examined the sociopolitical processes that shaped standardized innovation statistics and statistical manuals. Gault (2012, 2018, 2020) has been arguing for over a decade that standard statistical methods account for only a small portion of innovation activity and must be expanded beyond an exclusive focus on business. He points out that while innovation is broadly defined in the most recent edition of the OECD-Eurostat *Oslo Manual*, that definition is promptly 'put to one side to get on with innovation in the business sector' (Gault, 2020, p 102). And so, it is well established – by multiple scholars – that innovation statistics carry neoliberal politics. Here, I move from that *historiographic* style of number study to an *ethnographic* one (for this distinction, see Lippert, 2018, p 74, note 1). I consider how politics (and depoliticization) are enacted in the everyday use of conventional statistical tools and techniques.

As is the norm, I begin by presenting my descriptive statistics. Here we will see evidence that falsifies any claims against the existence of public innovation in goods. But we also see a lack of statistical significance. And so, I must do more. I follow the descriptive statistics with three sets of statistically significant results. These results follow the canonical progression of innovation theory I discussed in Chapter 2. There is a *linear* approach comparing innovation between public and private organizations, then network similarity tests on the *(chain) links* or *interactions* between organizations, and then statistical modelling of *system* fragmentation dynamics. I present each set of results in a conventionally acceptable style and format. Then, after each analytical progression, I break into an autoethnographic discussion on the meaning(lessness) of those results. The autoethnographic reflections culminate in a reframing of statistical practices through existential philosophy – specifically, Albert Camus' *The Myth of Sisyphus* (1955).

But first, let the statistical speak begin …

Descriptive statistics
Results

Following the survey methods described in Chapter 6, I produced a dataset covering 27 organizations engaged in the production and use of ocean science instrumentalities in Nova Scotia, Canada. This included 12 scientific instrumentality companies, ten PROs, and five public support organizations. Table 4 provides a summary of the product and process innovations reported by the 25 organizations where a key informant participated in the study.

All participating organizations were involved in the production of novel outputs (that is, outputs that were new to the world or new to their field, sector, or market) and had incorporated some process innovations over the past five years. Indeed, R&D intensity was high throughout the network: 44 per cent of the 1,783 employees were dedicated to research and/or development activities. The average R&D intensity of public support organizations was lower (16 per cent) than the R&D intensity of PROs (46 per cent) and companies (41 per cent).

I asked respondents to indicate the types of outputs produced by their organization over the past five years. All five product types were reported by a majority of respondents. This included 'instruments, machinery, and equipment' which were produced by 20 of the 25 responding organizations. It is interesting that all the companies, eight of the PROs, and one of the public support organizations engaged in the production of instruments, machinery, or equipment. Novelty levels were also high across all three types of organizations. All the PROs, nine of the companies, and three of the public support organizations reported introducing goods or services that were 'new to the world' over the past five years.

All responding organizations incorporated some degree of process innovation into their operations over the past five years. Nearly all

Table 4: Product and process innovations in Nova Scotia's ocean science instrumentalities innovation system

	PROs	Companies	Support organizations	Total
Number of organizations	10	12	5	25
Employees (full-time equivalents)	1,281	474	28	1,783
R&D intensity[1]	46%	41%	16%	44%
Product innovations[2]				
Percentage of organizations that produced:				
instruments, machinery or equipment	89%	100%	20%	80%
reports, information, documents, or manuscripts	100%	45%	80%	72%
computer software or datasets	78%	73%	60%	72%
education, training, or professional development	89%	73%	100%	84%
data collection, processing, or analysis services	100%	45%	60%	68%
Percentage of organizations introducing products that were:				
new to the organization	78%	73%	100%	80%
new to the field, sector, or market	89%	73%	60%	76%
new to the world	100%	82%	60%	84%
Process innovations[2]				
Percentage of organizations that introduced new:				
techniques or methods	100%	100%	80%	96%
machinery or equipment	100%	73%	80%	84%
software	100%	91%	80%	92%
Percentage of organizations introducing processes that were:				
new to the organization	89%	82%	60%	80%
new to the field, sector, or market	89%	64%	60%	72%
new to the world	100%	36%	20%	56%

Notes:

[1] R&D intensity is the proportion of total employees (full-time equivalents) devoted to research and/or development.

[2] Percentages based on 25 organizations, which excludes one non-responding PRO and one non-responding company.

organizations introduced new production techniques/methods and adopted new software. A sizeable majority (84 per cent) also reported introducing new machinery or equipment into their operations. There were fewer organizations involved in novel process innovations than in novel product innovations. Nonetheless, 56 per cent of organizations reported process innovations that were 'new to the world'.

The types of innovation and innovation novelty levels reported here confirm the high levels of innovation activity in this network. It is particularly important to note that PROs and public support organizations in this network all reported high levels of R&D intensity, product innovation, and process innovation. Most interestingly, innovative goods – instruments, machinery, or equipment – were produced, over the previous five years, by nine of the 14 public organizations in this study. Note that this finding alone runs counter to the widespread assumption – discussed in Chapter 1 – that innovation in goods is the exclusive domain of the private sector. These results are therefore revelatory in that they confirm the production of innovative technological goods by public organizations.

Significance?

'Revelatory' – what an understatement. I want to shout from the rafters about the importance of these numbers. They prove the existence of public innovation in goods! They contradict a widely held position about public sector innovation. And so, I think these numbers warrant a few adjectives. They deserve to have some rhetorical embellishment. They might even deserve to be described as 'highly significant'. But these words are policed in statistical discourse. Results can be significant or not. No descriptive adjectives are allowed. And the word 'significant' must be accompanied by a p-value. It cannot be used around purely descriptive counting. This means that my descriptive statistics lack any real description; they are merely a preamble to the statistical tests that will establish mathematical significance. At least that is the convention.

But surely positivist scholars still accept that even one observation of a 'black swan' will falsify a theory like 'all swans are white'. And make no mistake, this is the style of claim I am refuting with my descriptive data. Remember: 'technological innovations, especially goods, are the *exclusive domain* of the private sector' (Windrum and Koch, 2008, p 239, emphasis added). In *The Logic of Scientific Discovery*, Karl Popper railed against this kind of inductive 'all statement' (Popper, 2005, p 82). He was using an old metaphor (Taleb, 2010) – and doing so in a footnote, but his argument about black swans and falsifiability is legendary. In one of his appendices, Popper went on to argue that there is no need for probabilities (p-values) when testing statements with such certainty (Popper, 2005, p 378). Almost every (positivist) knows this principle (absent the details). And innovation studies are not immune. A famous paper by Philip Cooke asserts that 'although single cases should merely be heuristic rather than scientifically definitive, one alone is sufficient to refute conventional wisdom, rather as Karl Popper noted when a *black* swan was discovered in Australia' (2001, pp 945–6, emphasis in original). So, although innovation researchers are accustomed to complex econometrics, we need not count past one to observe dark innovation. I am no Karl Popper. But if falsifiability is indeed a positivist standard, then my point only needed one instance of public innovation in goods – one verifiable observation of one black swan. To my utter frustration, this was not enough.

Now that I have said this, I worry that the black swan metaphor muddies the waters. It implies that I am describing rare outliers – and so, it allows for dismissiveness. You see, an increasingly common use of the metaphor comes from Nassim Nicholas Taleb (2010). He capitalizes it as 'Black Swan' – to describe rare, high-impact, unpredictable events. But I was observing some things closer to black elephants than Black Swans. A black elephant is a phenomenon that 'either no one can see or chooses to ignore. Or, if its presence is recognized, no one is actually able to tackle it' (Sardar and Sweeney, 2016, p 9). Black elephants have high predictability, and yet they are often passed off as rare and random events (Gupta, 2009). And so, I cannot let the black swan metaphor go too far. There are nine public organizations in my dataset that produced new instruments, machinery, or equipment in the preceding five years. They are elephants in the room. It is not hard to predict their presence, but they are concealed by conventional wisdom, political belief, and measurement techniques. The mundane discourse and unreflexive standards of descriptive statistics makes them all too easy to ignore. And so, in the next section, I cave in to convention and start producing some *p*-values.

Locus of innovation

Results

As was noted in Chapter 2, prior research demonstrated that scientists, rather than private companies, are the locus of innovation for scientific instruments (von Hippel, 1976, 1988; Spital, 1979; Riggs and von Hippel, 1994). It follows that PROs – organizations that employ scientists and use scientific instrumentalities – will be the locus of innovation for a scientific instrumentalities innovation system. If we conceive of an innovation system as containing an interactive learning network, then we can use network analysis to assess the relative importance of different network positions.

The measure 'degree centrality' is typically interpreted as representing a node's importance or influence in a network (Borgatti et al, 2013). In one example from innovation studies, Takeda et al (2008) found that a multisector regional innovation system in Japan was characterized by several firms with high degree centrality that each served as hubs for geographical agglomerations of related firms. In another example, Gay and Dousset (2005) examined a network of biotechnology industry alliances and found that the most highly connected firms – those with high degree centrality – were the most likely to attract additional alliances over time. And so, my first hypothesis (H1) was that: *PROs have significantly greater average degree centrality than all other types of organizations in a scientific instrumentalities interactive learning network.*

I conducted a quadratic assignment procedure (QAP) (Hubert, 1987; Krackhardt, 1988; Martin, 1999) t-test to compare the degree centrality of public research organizations with the degree centrality of other organizations in the network: private companies and public support organizations. QAP is considered superior to ordinary linear regression for network analysis (Krackhardt, 1988). This resampling process takes observed data and randomly re-arranges the rows and columns of a dependent variable matrix. The relational structure of the dependent matrix is preserved, but it is no longer related to the independent variable matrix because observations have been reassigned to different nodes. This approach can be used to create a collection of observations that could have occurred at random. Properties of the observed data can then be compared against the properties of several thousand random permutations. The result of QAP is a permutation distribution that allows network analysis software to evaluate the statistical significance of observations: calculating the percent of random permutations that yield values greater or less than the observed values.

Based on the QAP t-test, the degree centrality scores for PROs ($M = 18.20$, $SD = 3.37$) were not significantly higher than the degree centrality scores for other organizations in the network ($M = 15.65$, $SD = 5.34$); $t(25) = 2.55$, $p = 0.11$. Hypothesis H1 was not supported. This result suggests that the slightly higher average degree centrality for PROs in this network could occur at random: a similar difference in means occurred in 11 per cent of 10,000 random permutations of the observed data.

In interpreting this result, it is important to note that the hypothesis was drawn from a literature on scientific instrumentality innovation that does not discuss public support organizations (see von Hippel, 1976, 1988; Spital, 1979; de Solla Price, 1984; Kline, 1985; Kline and Rosenberg, 1986; Rosenberg, 1992; Riggs and von Hippel, 1994; Gorm Hansen, 2011). Prior studies of scientific instrument innovation examined the relative importance of only two roles: 'users' and 'producers' (von Hippel, 1976, 1988; Spital, 1979; Riggs and von Hippel, 1994). These studies did not include any individuals or organizations that were similar to the public support organizations in Nova Scotia's ocean science instrumentality innovation system. It is possible that similar public support organizations did not exist at the time or in the context of prior research. Indeed, such organizations did not appear in the historical data for Nova Scotia's ocean science instrumentalities innovation system (see Chapters 3 and 4).

To further understand the impact of public support organizations on my results for H1, I conducted a post hoc hypothesis test (H1b). If this study had used a data sampling approach, post hoc hypothesis testing using classical statistical tests would be problematic; there would be a high risk of a type 1 error. However, there are fundamental differences between the assumptions underlying classical statistical tests of sample data and the assumptions underlying QAP hypothesis tests of whole network data (Krackhardt, 1988; Dekker et al, 2007; Borgatti et al, 2013). It is appropriate to state and test post hoc hypotheses in this study because the dataset includes the whole network population – not a sample, and because the significance of each result is evaluated using a new, randomly generated distribution of permuted observations – rather than an assumed normal distribution. Under these network analysis conditions, it is normal and appropriate to conduct post hoc tests (for example, Kilduff, 1992; Grosser et al, 2010; Soltis, 2012; Lopez-Kidwell, 2013; Tang et al, 2014) and to undertake exploratory data analysis (for example, Butts, 2008; de Nooy et al, 2011; Borgatti et al, 2013).

My post hoc hypothesis (H1b) was that public organizations have significantly greater average degree centrality than private companies in this network. I conducted a QAP t-test to compare the mean degree centrality of public organizations – PROs and support organizations – with the mean degree centrality of private companies. I found that the degree centrality scores for public organizations ($M = 18.47$, $SD = 2.87$) were significantly higher than the degree centrality scores for private companies ($M = 14.25$, $SD = 5.75$) in this network: $t(25) = 4.22$, $p = 0.02$. The post hoc hypothesis (H1b) was supported. This could suggest that public organizations – PROs and support organizations – are more important than private companies in the interactive learning network. The relatively lower degree centrality scores for private companies in this network is consistent with prior conclusions that private manufacturers are less important – not the 'locus' – for scientific instrument innovation (von Hippel, 1976, 1988; Spital, 1979; Riggs and von Hippel, 1994). The highest degree scores in this network are found among a combination of public organizations, including both PROs and public support organizations. This may suggest that public support organizations are an important extension of the scientific enterprise, even if their employees do not directly perform scientific investigations.

Because degree centrality is a common proxy for importance in a network (Gay and Dousset, 2005; Takeda et al, 2008; Borgatti et al, 2013), the foregoing is a common interpretation of differences in degree centrality. However, there is an alternative explanation that cannot be discounted: higher degree centrality scores could also suggest that public organizations in this system have a greater propensity to establish interactive learning relationships than private companies in this system. For now, a cautious interpretation of the results for H1(b) is that public organizations are more connected within this network than private companies. The relative importance of different organizations will be revisited in the results for H3.

Linearity?

Blah! What a convoluted way to say that Eric von Hippel (1976) was right. I put so much effort into understanding, executing, and describing these statistical methods. But even without understanding the nuances of QAP, there is a simple and clear (positivist) argument: this is the 'population' data for a network of 27 organizations. It is not a sample of organizations randomly selected from a larger population. Switching from the normal 'independent observations' logic into a 'network analysis' logic is like turning your brain inside out. Perhaps that is why I hear an audience in my head. The audience thinks me a statistical imposter. They are objecting to the post hoc test. They are telling me that t-tests are too simplistic. They are asking about control variables. And so, I have overexplained the QAP methods. I hedged the results with an alternate explanation. I even ran a second and unnecessary multivariate analysis just to be sure. I found that none of the usual control variables mattered. These additional results would have read:

> I conducted a multiple network regression to predict degree centrality from public/private organizational status, organizational age, size (in full-time employees), and R&D intensity. These variables did not significantly predict degree centrality, $F(4, 27) = 2.30$, $p = .091$, $R^2 = .17$. Only public/private organization status added significantly to the prediction, $p = .03$.

However, this paragraph and these results stayed in the 'pocket slides' at my thesis defence. They were not requested and so I did not present them. Questions about control variables had come up in a practice session, but not on the big day.

Overall, I was floored by how little reaction these 'locus of innovation' statistics produced. I did all this work and yet no one stopped me to say: 'Wait, why are you testing a linear model hypothesis and using linear statistics?' After all, I had parroted the argument that the linear model of innovation is out of date. This seems to confirm that Benoît Godin was right: the linear model of innovation persists today because it is 'entrenched' in statistics (2017, p 78). Standard innovation survey methods still carry linear model assumptions (Godin, 2017). Collecting data in this way makes it possible to perform linear statistical tests. And because such linear statistics are so 'ordinary' (pardon the pun), they allow the linear model of innovation to sidestep its persistent criticism. Ironically, this means that linear ideas about innovation are supported by circular reasoning. To break free, my next tests consider symbiosis and interaction (prepare to be bored by the most mundane of statistical backflips).

Chain links

Results

The literature on scientific instrumentality innovation discusses symbiotic relationships between those who produce science and those who produce scientific instrumentalities (de Solla Price, 1984; Rosenberg, 1992; Gorm Hansen, 2011). Therefore, relationships between PROs and instrumentality companies should include multiple concurrent types of interactive learning with knowledge flows in both directions. In network analysis terms, this means the relations should be multiplex and bidirectional. Stated as a hypothesis (H2), this means that *within a scientific instrumentalities interactive learning network, relations between PROs and private companies are multiplex and bidirectional.*

The ocean science instrumentality organizations I surveyed in Nova Scotia had a network density of 0.64, indicating that 64 per cent of the possible relations between any two organizations were present. Out of the 702 possible relations in this network, there are 240 possible relations between PROs and instrumentality companies. Interactions were reported for 124 of these dyadic pairs. Seventy-four of these interactions were multiplex. Ninety-two relations were bidirectional. Seventy relations were both multiplex and bidirectional.

I calculated a Jaccard similarity coefficient to assess the degree to which the set of relationships between PROs and instrumentality companies intersected with the set of multiplex and bidirectional PRO-company relations. For this test, the Jaccard coefficient was more appropriate than a Pearson correlation coefficient because the data are binary (Hanneman and Riddle, 2005). The Jaccard coefficient is an index of the similarity between two sets of binary values. The hypothesis was focused on the composition of PRO-company relations, so the test was conducted using only the data on PRO-company dyads. In other words, support organizations were not included in this analysis, nor were PRO-PRO and Company-Company relations. The results of the test were assessed for significance using the QAP with 10,000 permutations. The distribution of similarities for the 10,000 random permutations ranged from 4 per cent to 54 per cent ($M = 23.2$ per cent $SD = 6.3$ per cent). I found a significant similarity between the two sets of relations: $J = 0.56$, $n = 124$, $p < 0.001$. The majority (56 per cent) of observed relationships between PROs and instrumentality companies were multiplex and bidirectional. Hypothesis H2 was supported. This result affirms prior discussion about the nature of interactive learning in scientific instrumentality innovation (see de Solla Price, 1984; Rosenberg, 1992; Gorm Hansen, 2011).

Triviality?

'This result affirms prior discussion' – what a feeble attempt to justify inane details. In my thesis, I went even further. I dedicated pages and pages of analytical discussion to showing that the transfer of equipment and technical services is a critical 'channel' of interactive learning between PROs and private instrumentality companies. But none of the numbers in the previous section or in the thesis add substantively to our understanding of scientific instrumentality innovation. This is merely quantification of insights that were established many years ago. Yet, the numbers seem to add value. They suggest greater rigour than the previous qualitative studies. The numbers suggest greater certainty. They seem more definitive. But make no mistake: there is nothing innovative about these innovation statistics. They are an extraordinarily incremental contribution. They are rigour to the point of rigor mortis.

Why couldn't I admit that these results are trivial? Because for me – their author – these numbers were both a fait accompli and a major feat. I knew what they would say. But I was also tremendously proud to have produced them. These data represent months of effort. It was like solving a complex puzzle: finishing it made me feel clever and accomplished. I impressed myself and I hoped this work might also impress others. I felt like a *real* social scientist because I was able to produce *really* complex statistics. But in so doing, I made the results inaccessible to anyone who might use them for shaping policy or practice.

Policy makers would be better advised to read one of the qualitative studies anyway. De Solla Price (1984) doesn't bore you with unintelligible mathematics. But short of the numbers, work like his feels less certain, less dependable. Ironically, policy makers are more likely to respect my quantification, but less likely to understand it. I made these ideas trivial through mathematics. This is not unlike Saifer and Dacin's observation that 'the overproduction of data doesn't lead to more knowledge, but rather greater levels of organizational ignorance' (2021, p 627). In the 1960s, Ernest Becker warned that research was 'becoming mired in data and devoted to triviality' (Becker, 1968, p xiii). More recently, leading autoethnography scholar Art Bochner has warned that this 'devotion to triviality can lead to alienation' (2016, p 51). Nonetheless, I will now try to eke out some meaningful impact. I now turn to a statistical test that aims to mirror a real-world innovation system dynamic (but let's not forget to mute the politics).

System dynamics

Results

In Chapter 3, I described a major innovation system dynamic that occurred five years before my data collection: substantive reductions in funding for public science across the country, and particularly in ocean science (Turner, 2013). This was part of a broader decline in public science globally that will have 'long term adverse consequences' (Archibugi and Filippetti, 2018, p 108) for innovation and development. Here in Nova Scotia, reduced funding for ocean science stood in contrast to increased emphasis on ocean technology development. Indeed, my stories in Chapter 4 suggested that Nova Scotia's ocean science and technology innovation system might be structurally dependent on public research organizations as its 'anchor tenants' (Agrawal and Cockburn, 2003; Niosi and Zhegu, 2005; Niosi and Zhegu, 2010).

In graph theory, the structural dependence of a network on certain nodes is referred to as 'robustness' (Callaway et al, 2000; Barabási, 2013). A network's robustness is a function of how well it remains connected when individual nodes or edges are removed (Borgatti et al, 2013). A network is said to be highly robust when a large number of nodes or edges need to be removed before the network begins to fragment into many small components (Borgatti et al, 2013). Robustness has mostly been qualitatively explored in innovation studies. Some have suggested that Silicon Valley's present-day innovation system is highly susceptible – not robust – to the loss of venture capital firms (Ferrary and Granovetter, 2009). Others have suggested that Boston's biotech innovation system was not robust to the removal of PROs in the late 1980s (Powell et al, 2012).

The dynamic effect underlying network robustness is fragmentation. In a network with no fragmentation, all nodes are members of one component – no individual nodes are isolated from the group, and no small groups of nodes are disconnected from the main component. When there is no fragmentation present, any node in a network can reach any other node by working through its neighbours. For an innovation system network, this could mean that knowledge and learning can flow efficiently and effectively.

Stephen Borgatti (2006) identified several ways to measure network fragmentation. For all these measures, a network becomes fully fragmented ($F = 1$) when all nodes are disconnected from one another. Fragmentation measures differ in the ways that they account for degrees of fragmentation. The simplest approach is to count the number of components – or groups of nodes – in a network and then divide them by the total number of nodes. Using this measurement technique, Calignano, Fitjar, and Kogler (2018) observed that the aerospace cluster in Apulia, Italy was highly fragmented in a static sense. The whole network's degree of fragmentation was measured at two separate points in time, without modelling perturbations between time periods.

Borgatti (2006) argues that we should go further than measuring fragmentation as a static state. He suggests that we should account for the impact on network structure that occurs when nodes are lost: the loss of a well-positioned node, one with high degree or betweenness centrality, for example, can have greater implications for the functioning of a network than the loss of a peripheral node (Borgatti, 2006). To calculate the impact of node loss on network structure, Borgatti (2006) considers the reciprocal distance between nodes. In other words, he measures the degree to which any pair of nodes in a network can reach one another via connections with their neighbours. He calls this 'distance weighted fragmentation' (^{D}F) (Borgatti, 2006). In practical terms, reachability is the number of edges that a piece of knowledge must traverse to find its way from one organization to another in an interactive learning network. After incorporating reachability into a measure of fragmentation, Borgatti (2006) gives us this equation:

$$^{D}F = 1 - \frac{2\sum_{i>j}\frac{1}{d_{ij}}}{n(n-1)}$$

Here, i and j are nodes in a network, d_{ij} is the geodesic distance between those nodes, and n is the total number of nodes in the network. The numerator incorporates a reciprocal of the distance between nodes. For nodes that cannot reach one another – in other words, distance is infinite – the reciprocal distance is zero. Distance-weighted fragmentation has a lower limit of zero when every pair of nodes is adjacent to every other pair. It has an upper limit where every node is an isolate. For my purposes, distance-weighted fragmentation is useful because it can be a node-level measure: the change in ^{D}F of the network can be calculated after removal of any individual node. This concept of distance-weighted fragmentation allows me to hypothesize (H3) that: *removing individual PROs from a scientific instrumentalities interactive learning network results in significantly greater distance-weighted fragmentation than removing other types of organizations.*

I conducted a QAP *t*-test to compare the mean change in ^{D}F after removal of a PRO with the mean change in ^{D}F after removal of other organizations in the network (private companies and support organizations). The fragmentation scores for PROs ($M=0.002$, $SD=0.005$) were not significantly greater than the fragmentation scores for other organizations in the network ($M=-0.001$, $SD=0.008$): $t(25) = 0.004$, $p = 0.11$. Hypothesis H3 was not supported. This result suggests that the larger average fragmentation scores that were observed for PROs in this network could occur at random: differences that were the same as or greater than the observed difference occurred in 11 per cent of 10,000 random permutations of the observed data.

The result for this test is like the result for the test of degree centrality scores (H1). As with hypothesis H1, I formed a post hoc hypothesis to account for the presence of public support organizations in the data (H3b): *removing individual public organizations from a scientific instrumentalities interactive learning network results in significantly greater distance-weighted fragmentation than removing private companies.*

I conducted a second QAP *t*-test to compare the mean fragmentation scores for public organizations (PROs and public support organizations) with those for private companies. The fragmentation scores for public organizations ($M=0.003$, $SD=0.004$) were significant greater than the fragmentation scores for private companies ($M=-0.004$, $SD=0.009$): $t(25) = 0.006$, $p = 0.013$. The post hoc hypothesis (H3b) was supported. This result suggests that, on average, this innovation system would become more fragmented following the loss of a public organization than it would become following the loss of a private company.

Attack!

Convention is clearly the enemy of antagonism. Here, as in my PhD thesis, I have provided a conventional description of my system fragmentation analysis. In fact, the four results sections in this chapter all follow the conventions for presenting statistical results set forth by the APA. Although I used UCInet to produce the statistics, I paid for a subscription to Laerd Statistics and followed its templates for converting statistical results from SPSS (Statistical Package for the Social Sciences) into the writing style required by the American Psychological Association (APA). To write the 'I conducted ...' paragraphs in this chapter, I simply filled in the blanks in the relevant templates. The outcome of these templates is predictable. It is an understated, technocratic description of a relatively complex statistical analysis. The writing conventions give a sense of rationality and objectivity. They depoliticize the discussion. And yet, I was trying to mount a major counteroffensive in the Canadian 'War on Science' (Turner, 2013).

In the final months of my PhD studies, it became clear that these statistical results were acceptable, but their politics were not. I will not recount the micropolitics that played out. But the big 'P' politics are critical to my arguments in this book. In presenting the analysis in the previous section, I dropped any sense of its political motivation. Following that convention was necessary to complete my PhD. But Godin (2005) demonstrates that statistics on science and technology are first political, before they are ever (re)presented as objective. Once wrapped in 'the optics of neutrality' (Saifer and Dacin, 2021, p 632), data 'is often mobilized to obscure or depoliticize' (2021, p 625). And here, depoliticization refers to 'the processes through which societal issues (represented via data points) become decoupled from their political and structural roots, and through which solutions are conceptualized within a narrow frame that is seen as "beyond" or "post"-politics' (Saifer and Dacin, 2021, p 632). Adam Saifer and Tina Dacin (2021) argue that organization studies ought to consider people's aesthetic, emotional, and discursive engagement with numbers and statistics (Saifer and Dacin, 2021). Meanwhile, Ingmar Lippert and Helen Verran call for researchers 'to employ, further develop, interrogate STS number analytics and study numbers' (2018, p 9). Only the mainstream positivists remain committed to the depoliticization of numbers.

To meet disciplinary expectations, my results needed to appear purely rational. The norms of innovation studies allow a dispassionate simulation of innovation system robustness. This simulation can be presented as a valuable methodological innovation. But it is entirely another thing to use numbers in a passionate critique of neoliberal dogma. And so, in concluding my thesis, I carefully and cautiously described the implications this way:

> Five years ago, substantial federal cuts were made in ocean science across Canada (Bailey et al, 2016; Turner, 2013) at the same time as regional policy networks were prioritizing investments in ocean technology innovation via industrial policy (Government of Nova Scotia, 2012; Greater Halifax Partnership, 2012). Ocean science and ocean industry policies were moving in opposite directions. My results suggest that this disconnect may have been problematic because, in the interactive learning network that I observed, the loss of a public organization would cause greater fragmentation to the network – on average – than the loss of a private company. This suggests that the innovation system may be structurally dependent upon public organizations. Furthermore, I found that the majority of interactive learning relationships between PROs and private companies in this network were symbiotic. This suggests that it may be important to connect public policies in support of private companies in this system (i.e., industrial policies) with policies that affect PROs and public support organizations (i.e., science policies).

Notice the muted phrases like 'may have been problematic' and 'this suggests'. This soft language leaves room for neoliberalization: these results can be read as an indication that ocean technology innovation in Nova Scotia should become less dependent upon public organizations. My argument would have been the opposite: ocean science is a fundamental public good.

So do not assume from the muted tone of my statistical speak that I wasn't angry about the cuts to public ocean science. Do not assume that I wasn't also angry about tempering my language. The passion and politics were never absent from my numbers; they were just thought to be incompatible, by convention. My 'attack' became blunted. Ironically, there is research demonstrating that more emotion (not less) would have 'energized' engagement with the data (Saifer and Dacin, 2021). A passionately mobilized critique of federal policy may have registered in the public discourse. But I became tired of policing myself through the enumeration. And by the time I had finished the work, a new federal government had already begun reversing the cuts to science – while even more dramatically increasing its financial support for private sector ocean technology companies.

Significant but meaningless

In this chapter, I produced and critiqued four sets of statistics corresponding to three perspectives in innovation theory. First, I presented descriptive statistics that should have been sufficient evidence of the 'black elephant' that is public innovation in goods. But alas, this counting was not statistically 'significant'. Next, I tested insights from linear model studies of scientific instrument innovation. I found that the locus of scientific instrumentality innovation rests with public sector organizations. However, my *representative* – but fictional – audience got caught up in a critique of the number-crunching details. It was less obvious that the whole exercise was stuck in a vicious cycle of linear assumptions, models, and statistics. Then, I tested old insights about the symbiotic and interactive relations between scientists and instrument manufacturers. I felt clever in enumerating those old insights. The results appeared more dependable than past research, but were completely trivial. Finally, I strived for 'real world' impact by mathematically testing a misguided public policy. The numbers supported my position, but that position was undermined by representational conventions. At each of these three stages, I was pushed forward by enthusiasm, optimism, and the intellectual challenge. Then, when the work was done, I was deflated by anger, frustration, and disappointment. This is how statistics held their sway for so long in my life. These tools help me feel clever, accomplished, and accepted (at the disciplinary 'convention' – or 'gathering'). I kept returning to statistics for these reasons and they kept letting me down. Each time I write statistical results, their incremental futility surfaces. These are the moments when the whole statistical exercise feels Sisyphean. These are the moments when the statistical work – work that is so valued in innovation studies – is revealed to be significant, but meaningless.

Through the myth of Sisyphus, we can begin to see the absurdity of producing innovation statistics, and other social science quantification, and life in general. In existential philosophy, the absurdity of life is that we aspire to find meaning, but the universe ultimately leaves us without answers. This can lead to nihilism: the position that life is meaninglessness and devoid of any objectively unifying truth or values. Albert Camus (1955) argued that even if we accept this as our reality, it does not mean we should give up. To make his point, Camus imagined Sisyphus as an absurd hero. He imagined that Sisyphus might be happy whenever his boulder rolls back to the bottom of the hill. Those moments of happiness are possible because Sisyphus is aware of his absurd circumstance:

> Sisyphus, proletarian of the gods, powerless and rebellious, knows the whole extent of his wretched condition; it is what he thinks of during his descent. The lucidity that was to constitute his torture at the same time crowns his victory. There is no fate that cannot be surmounted by scorn. (Camus, 1955, p 109)

Here, Camus is arguing that we cannot find meaning in life by pretending that all will be well tomorrow or by hoping that some God will eventually save us. Nor can we simply give up: in the absurdity of life, 'suicide is not legitimate' (Camus, 1955, p 8). Instead, Camus provides a book-length argument for consciousness and 'revolt' (see especially, Camus, 1955, pp 53–5). And in my own little way, I have carried that thought into the realm of innovation statistics.

Rather than giving up on statistics, I have embraced the absurdity of the exercise. I have not retreated from this absurdity through any 'philosophical suicide' (Camus, 1955, p 32) – that is, the kind of *escape to certainty* where I might place my faith in some other universal ideal. That would be no less absurd than placing my faith in positivism, neoliberalism, or numbers. After all, statistics are a 'desecularized' religion – part of the scientific substitute for God (Gephart, 2006, p 426). This chapter has been a smirk at the absurdity of the statistical religion – an affirmation of my own experience, voice, and freedom. Importantly, I have not claimed that the numbers I produced were disconnected from reality – that is the kind of 'absurdity' that logical positivist philosophers get worked up over. Rather, I have tried to share my experience of personal alienation-through-statistics. I hope that this has 'resonance' (Ellis, 2004, p 22) for other *recovering positivists* who are similarly tired of fragmenting their identities to please the statistical gods. We need not succumb to this discipline.

Autoethnography helped me work through the alienation of statistics and produce a contribution here that I consider meaningful. Making meaning for self and others is the whole point of autoethnography. Art Bochner has described it as 'an expression of the desire to turn social science inquiry into a non-alienating practice' (2016, p 53). In this way, 'it's a response to an existential crisis – a desire to do meaningful work and lead a meaningful life' (Bochner, 2016). Like me, Bochner read Camus in high school and agrees with the Sisyphean lesson: 'it is what we create ourselves, what we experience and do, that gives meaning to our lives' (Bochner, 2016, p 50). However, Bochner would contrast – not combine – statistics and autoethnography. He explains that 'whereas empiricist social science fuels an appetite for abstraction, facts, and control, autoethnography feeds a hunger for details, meanings, and peace of mind' (Bochner, 2016, p 53). I agree with him in the end, but it was my journey through auto-ethnostatistics that brought me this peace of mind.

I submit that this is the value of fusing autoethnography and ethnostatistics. This fusion can help us create (our own) meaning from 'inside' statistical tools, techniques, and practices. This idiographic and 'situated' meaning is a revolt against the absurd. In response to Lippert's call for more 'tools to open up numbers and calculations' (2018, p 53), my autoethnography of statistics is also an alternative and/or addendum to Gephart's (1988) ethnostatistics, Callon and Law's (2005) qualculation, Verran's (2001) 'ontologizing troubles' (Lippert, 2018), and B. T. Lawson's (2023) 'life of a number approach'. These and many forms of 'number study' involve the analysis of other people's enumeration – assessing the counting within other people's knowledge claims. Verran (2001) seeks some redress from this when she decomposes her own analysis. But the 'auto-' that I have invoked in this chapter precludes us from taking any God-like position in the first place – it inhibits what Haraway (1988) called 'the God trick'. Instead, autoethnography pushes us through the discomfort of our own experiences and demands radically reflexive authenticity (Ellis, 2004). In my next and final chapter, I will argue that this kind of reflexivity must be the container for any dark innovation toolkit.

8

After Observation

Entanglements

I began this book with Ben Martin's (2016, 2013) 'dark matter' analogy. I agreed that important phenomena are absent from innovation studies and yet I disagreed about the reasons. For Martin, the dark innovation challenge is about deficiencies in measurement. He was calling for new social science instrumentalities analogous to the supercollider technologies that he and Ben Irvine wrote about at the Conseil Européen pour la Recherche Nucléaire (CERN) (Irvine and Martin, 1984a; Martin and Irvine, 1984a, 1984b). From that perspective, researchers seem to need more ingenious measurement instruments to *push* the limits of our knowledge about innovation. Conversely, this book has been about *surfacing* the limits of our knowledge – inquiring into some assumptions that are embedded in conventional methods. In each chapter, I have attempted to fracture a methodological norm that is taken for granted in mainstream innovation studies. Through these cracks in the disciplinary matrix, we could see that methods conceal as much as they reveal. Empirically, I was interested in the ways that methods can conceal or reveal public innovation in goods. But that specific instance of dark innovation was a means to an end. The end is a much different dark innovation challenge than the one set out by Martin (2013, 2016). It is not so much about where or what we might observe next; instead, it is a call to critically examine *how* we understand innovation: to deconstruct the instrumentalities used in innovation research.

Interestingly, this is exactly the direction in which the dark matter analogy takes us (if we read the philosophy and sociology of physics). Because they recognize the complete mediating role of their instrument configurations, high energy physicists – like those at CERN – are obsessed with understanding their devices and surrounding assumptions (Knorr Cetina, 1999, p 56). In her ethnography at CERN, Karin Knorr Cetina noticed that 'more time in an experiment is spent on designing, making, and installing its own components, and in particular on examining every aspect of their working, than on handling the data' (1999, p 56). She stated that 'the behaviour of this apparatus, its performance, blemishes, and ailments are not self evident to the physicists. These features must be learned, and the project of understanding the behaviour of the detector spells this out' (1999, p 56). In other words, Knorr Cetina observed physicists acting as if all their instruments have poorly understood limitations. In their attempts to understand quantum mechanics, dark matter, and so on, these physicists were investing most of their time and energy in understanding their instrumentalities.

Remarkably, the scientists told Knorr Cetina that their measurements would be 'meaningless' (Knorr Cetina, 1999, pp 52–5) without this deep understanding of the instrumentalities. Here, as in Chapter 7, no one is suggesting that observations or measurements are completely disconnected from a material reality; instead, this is an assertion that measurements and instrumentalities are inseparable. Knorr Cetina explains that

> in many fields, measurements, provided they are properly performed and safeguarded by experimenters, count as evidence. They are considered capable of proving or disproving theories, of suggesting new phenomena, of representing more or less interesting – and more or less publishable – 'results' ... In high energy collider physics, however, measurements appear to be defined more by their imperfections and shortcomings than by anything they can do. It is as if high energy physicists recognized all the problems with measurements that philosophers and other analysts of scientific procedures occasionally investigate. (Knorr Cetina, 1999, p 53)

These scientists were not attempting to fix the problems of measurement. They were not trying to eliminate the 'errors' or 'biases' from their instruments. They were trying to account for the various ways in which their instruments mediate their observations of the universe.

Karen Barad (2007) takes this point further in *Meeting the Universe Halfway* – a book that fuses insights from quantum physics and queer feminist theory. Barad is both a bona fide physicist and a highly cited philosopher of science. In their groundbreaking book, Barad asserts that 'one must inquire into the material specificities of the apparatuses that help constitute objects and subjects' (Barad, 2007, p 115). A rough understanding of quantum physics might lead us to agree: 'Of course! Didn't someone show that photons become fixed as either waves or particles, depending on how we measure them?' But this observation-is-everything argument harkens back to one of Heisenberg's early formulations of the uncertainty principle – one he knew to be wrong (Barad, 2007). There is a risk that we social scientists might follow that argument back to the conclusion that all knowledge is relative. What Barad tells us, instead, is that 'there is something fundamental about the nature of measurement interactions such that, given a particular measuring apparatus, certain properties become determinant, while others are specifically excluded' (Barad, 2007, p 89). It is not that measurement is entirely meaningless – that is, detached from reality. Rather, observations and instrumentalities are entangled. Barad sums this up for all researchers by saying 'our knowledge making practices are socio-material enactments that contribute to, and are part of, the phenomena we describe' (Barad, 2007, p 113). Here, the physics of entanglement resemble the philosophies of knowledge; both tell us we must be attentive to the configuration of scientific practices.

For this reason, this chapter will not include a list of the 'most promising' places to find dark innovation next, or a list of the 'best' instrumentalities for studying dark innovation. Instead, it will conclude my argument for open inquiry into the instrumentalities of innovation research. My point of departure for this concluding chapter has been the idea that observations and instrumentalities are entangled. Next, this leads to a brief review of the observations and 'equipment list' contained in this book. Then, to reaffirm that all methods create absences, I share a brief confessional account on a highly important ocean innovation I overlooked. That oversight brings me deeper into the work of Barad (2007) and Donna Harraway (1988, 1990) to situate 'the researcher' in any dark innovation toolkit.

Instrumentalities

It would be fair to say that I chose an easy path into questions about dark innovation. Neoliberalism and innovation are *closely coupled* concepts. Langdon Winner has argued that the 'cult of innovation' is 'the jewel in the crown of neoliberalism' (2018, p 67). There is also already a relatively well-developed critique of neoliberal ideology in innovation studies (see Fløysand and Jakobsen, 2011; Gallouj and Zanfei, 2013; Cruz et al, 2015; Cooke, 2016; Lundvall, 2016; Godin, 2017; Pfotenhauer and Juhl, 2017; Winner, 2018). Yet, my experience has been that calling out these politics remains 'taboo'. The 'epistemic culture' (Knorr Cetina, 1999) of innovation studies not only discourages critique of neoliberal perspectives, but also inhibits alternative knowledge. For example, we have now seen how some artefacts and practices of mainstream innovation research inhibit direct knowledge about public innovation in goods. Benoît Godin said that 'the persistence of the market-first perspective speaks more about the values of the scholars promoting it than to its contribution to understanding technological innovation' (2017, p 125). But I have been hesitant to accuse any individual of explicit neoliberal bias. From Chapter 1 onwards, I have argued that neoliberalization is a disciplinary achievement, and I have called for change in the instrumentalities of innovation studies.

In Chapter 2, I approached theoretical models as especially powerful 'noncorporeal' (Hartt, 2013, 2019) tools or instruments. Important innovation scholars observed the same phenomena (scientific instrument innovation) to behave very differently through different theoretical assumptions, and this led to quite different theoretical models. I noted how these different models share a focus on firms and markets. And this is not the only way in which theoretical models mediate knowledge of innovation. Godin pointed out that innovation models generally exclude 'human and social needs' (2017, p 125). He lamented 'the lack of reflexivity on models of innovation (models have a history that is too often forgotten)' (2017, p 3). More research is needed.

In Chapters 3 and 4, I engaged with the instrumentalities of history. First, I suggested that we cannot take the idea of 'context' for granted. We must consider how, why, and by whom stories of the industrial/technological past are written. This includes the stories we write in our research. In Chapters 3 and 4, I had to make choices about what to include and exclude. Also, only certain material traces were available (because the past is inaccessible to us). However, I made my intentions clear, cited the traces of the past I deployed, presented plausible arguments, and remained open to other possibilities. Indeed, Chapter 4 was about opening up possibilities for knowing and telling stories of innovation. I confronted the metanarrative of neoliberalism with three short stories that focused on public organizations and did not end in 'market' resolution. This recharacterized public research organizations as active agents and private companies as supportive quartermasters. These stories were written against the grain of narrative neoliberalization, and there are other innovation metanarratives worth deconstructing in this way in future research, such as 'evolution' and 'progress'.

Chapter 5 was focused on the most popular set of classification tools in innovation studies: the taxonomies of innovation. Beyond the well-criticized issues of industrial classification, I examined the instrumentality that makes the whole taxonomic puzzle possible: the organization-as-organism metaphor. I argued that this biological metaphor was a key inscription point for conservative, neoliberal ideas about innovation. Old ideas about politics and economy persist through Pavitt's taxonomy and its heirs. These taxonomic ideas continue to shape the organizations and industries we value. They constrain our ability to notice things that do not quite fit, such as public and nonmarket innovations. At one extreme, some criminal and terrorist activities are enacted in 'innovative' ways to ensure that no one will be able to 'pin down' the organization/organism. At another extreme, we should note that humans are not the only biological organisms involved in organizing for innovation (see O'Doherty, 2023). Overall, much more work is needed on questions of classification and valuation in innovation studies. Even in biology, the metaphors and assumptions that drive taxonomic classification are being debated.

I considered the possibilities for alternative innovation metaphors in Chapter 6. There, my focus was on the normal topological assumptions made when framing innovation systems and collecting survey data. I examined some specific ways in which the default 'regional' topology requires 'surveyable' boundaries. Some things must be excluded to draw boundaries around a regional volume. But different topological metaphors – network/rhizome, fluid, and fire – can also bring different understandings of innovation into focus, and this was evident in the margins, gaps, and shadows of my ocean science instrumentality data. I argued that alternative topologies are critical for observing dark innovation – just as some physicists have argued that dark matter and dark energy might not be observable in Euclidean space. Our default topologies are Euro-colonial (Law, 1999) and will not be able to help us see past the 'Western bias' (Chaturvedi, 2023) in innovation studies. Future research might turn to new metaphors, like 'hyperobject' (Rehn and Örtenblad, 2023). Or future research might aim for the 'interruption of topology' by advancing Damian O'Doherty's ideas about 'topology of method' (2013, p 213).

Although Chapter 7 used personal narrative (auto-), my inquiry was into the cultural experience (ethnography) of producing innovation statistics. It was my expression of existential 'revolt' against statistical norms. I recounted the Sisyphean task of producing statistical evidence that might have some 'meaningful' scholarly and policy impact. In the end, I found myself alienated from statistical practices that produced linearity, triviality, and insignificance (in the *true*, nonstatistical sense). Turning to autoethnography allowed me to make meaning from the frustrations of this work. Given the gendered and racialized nature of datafication (D'Ignazio and Klein, 2020), I believe that we need more autoethnographies about innovation numbers and statistics from non-White, nonmale scholars (for example, Liu and Pechenkina, 2019). There is also much more unrealized potential for autoethnography (and existential philosophy) in critical studies of innovation.

In this way, Chapter 7 was also an echo of the ideas from Karl Weick that I introduced at the end of Chapter 2. Weick (1996) noted that tools can be cultural and personal. Reflecting on his study of the Mann Gulch firefighting disaster, he advised (management) academics that disciplinary affinities can cause us to hold some tools too tightly: 'the fusion of tools with group membership makes it hard for firefighters to consider tools as something apart from themselves that can be discarded, just as it makes it hard for scholars to consider concepts as something apart from themselves' (Weick, 1996, p 312). Concepts and other instrumentalities are entangled in any disciplinary construction. And yet, the dark innovation challenge cannot be resolved by swapping out one set of tools for another. To illustrate, let me recall an email that forced me to drop the tools I used in this book.

Absences

It was a brief message. I had recently 'finished' my research at the Nova Scotia Public Archives. Archivist Rosemary Barbour was following up to let me know that the Elizabeth Mann Borgese fonds were now available at the Dalhousie University Archives. Unfortunately, I had no idea who Rosemary was writing me about or why this might be important. I had no memory of encountering Elizabeth Mann Borgese in the traces of past ocean science and technology efforts in Nova Scotia. So, I promptly googled the name. I was floored by the results. I had missed Nova Scotia's most influential oceans innovator and her radically disruptive innovation.

Elisabeth Mann Borgese – the 'First Lady of the Ocean' (Inglott, 2004) – was a founding member of the Club of Rome, a diplomat, activist, environmentalist, and law professor at Dalhousie University. She organized the first annual *Pacem in Maribus* conference at Malta in 1970 (Borgese, 1973). It became a series of conferences that shaped the United Nations Convention on the Law of the Sea (UNCLOS). In a book about Mann Borgese's life, legacy, and the UNCLOS process, Tirza Meyer explains that:

> Elisabeth Mann Borgese wanted a new order for the oceans at a time when there were almost no rules governing them. A time when the extent of a nation state's sovereignty over its coastal waters was still measured by the distance of a cannon shot (about four nautical miles) and fishing and transport rights were negotiated bilaterally. In fact, Mann Borgese's ambitions went beyond even this – she wanted a fairer system of governance not just for the oceans but for the entire world. (Meyer, 2022, pp 3–4)

While it is not possible to draw a straight line between Mann Borgese and the words contained in the UNCLOS (Meyer, 2022), there is little doubt that her diplomatic efforts shaped the international governance of all human relations with the ocean. Her work was what the OECD (2005) might label 'organizational innovation'; it disrupted the way we organize ocean governance. As a result, at the time of her death in 2002 (aged 83), she had been awarded the Order of Canada, the German Order of Merit, the Order of Columbia, the Austrian Metal of High Merit, the Friendship Prize from the People's Republic of China, the United Nation's Sasakawa International Environment Prize, and five honorary doctorates (including one from my alma mater, Mount Saint Vincent University) (International Ocean Institute, n.d.). Her legacy continues today through the International

Ocean Institute, an independent NGO she founded in 1972 to develop capacities in international governance for ocean sustainability (International Ocean Institute, n.d.). I could not understand how I missed this woman and her work.

Once I went looking, Elizabeth Mann Borgese was easy to find. For example, it took very little time to walk across my campus, pull one of her books from the library stacks, and locate her ideas about ocean technology in the index. It seemed that Mann Borgese (1998) had anticipated the way I would emphasize technoscientific innovations over the sociopolitical ones she had advanced. She said

> it is the nature of the oceans that pushes science and technology into the foreground. Without marine science and technology we would be blatantly unable to explore, exploit, manage, and conserve marine resources or to navigate safely or to protect our coasts. And it is the nature of the marine environment that forces us to recognize that this science must be interdisciplinary ... and that it must be international. (Mann Borgese, 1998, p 114)

But on the same page she also warned that 'there is indeed a danger inherent in the strong emphasis on science and technology. In ocean governance, given the imbalance between "North" and "South," this emphasis could reinforce the dominance of the North' (Mann Borgese, 1998, p 114). And elsewhere in the same book, she explored Ghandi's ideas about 'appropriate technology' for the ocean – technology that meets human needs and does 'no harm to the body, mind, or soul' (Mann Borgese, 1998, p 97).

She was right, of course. We get carried away by the promise of technoscience for development. That discourse privileges the Global North. It assumes that all innovation is good. To produce the 'public goods' of peace and environmental sustainability, we need innovation in global ocean governance. In my prior career, I knew that good governance must always come before flashy technoindustry. And yet, in my research, I had done exactly as Elizabeth Mann Borgese worried we might all do: I brought technoscience into sharp focus and missed the immense importance of activism, diplomacy, and advocacy.

Some might say that I was in the dark about Mann Borgese because there were archival silences – that is, my awareness of her work was mediated by the politics of the archive and its traces (for example, Corrigan and Mills, 2012). In other words, there was missing data, or I had not dug deeply enough. Others might say that I was using the wrong tools – that I approached the archives with an ANT toolkit that is insufficiently critical (for example, Whelan, 2001; de la Bellacasa, 2010). Still others might say that my ignorance towards Mann Borgese and her work points away from dark innovation and towards an entirely different problem in innovation studies. Ben Martin (2013, 2016) would call this the 'boys' toys' challenge – that is, the challenge of gender bias in innovation studies.

While Martin outlines 'dark innovation' as a challenge of unknown absences, he sees 'boys' toys' as a challenge of excessive presences. He points out that the journal *Research Policy* is overwhelmingly focused on computers, cars, televisions, and various electronics (Martin, 2013, 2016). He attributes this 'skew' to two factors: '(i) a high proportion of researchers in the field are men; and (ii) researchers are likely to focus their empirical work on an area they feel passionate about' (Martin, 2013, p 172). His solution is also twofold: to increase the proportion of female researchers in the field and to produce research on innovations 'that have freed women from the domestic drudgery of being "housewives"' (Martin, 2013, p 172), such as refrigerators, microwaves, and washing machines (Martin, 2016). The first of these is a proposal with which I can agree, while the second is a proposal that I find highly problematic (not only because I am surprised by such gendered ideas about housework). Here, Martin fails to notice decades of research on gender and technoscience: research that led me to the conclusion that 'we are the tools'.

Tools

I stood up at the 2019 *International Critical Management Studies Conference* and called myself a 'tool' for overlooking Elizabeth Mann Borgese (MacNeil, 2019). On the one hand, I was invoking the vulgar insult I might have received in high school after making such a stupid mistake. On the other hand, I was trying to throw myself in with the whole set of instrumentalities that had surfaced *public innovation in goods* while absenting Mann Borgese's *innovation for the public good*. I was, of course, agreeing with Anne-Jorunn Berg and Merete Lie that 'artifacts do have gender and gender politics in the sense that they are designed and used in gendered contexts' (1995, p 347). The technologies of ocean science – indeed, all scientific instruments – tend to be masculinized. I was also attempting to acknowledge my 'geek masculine' identity work (Bell, 2013; Morgan, 2014) and surface 'the male pleasures that are made in the knowing and telling of machines' (Law, 1998, p 45). But my experience was also unlike the one that John Law describes in his rambling essay about the 'machinic pleasures' that 'interpellated' him into a study of military technologies (Law, 1998). Missing Mann Borgese was not as crude and simple as 'men love machines' (and 'women care for nature/others'). I agree with Sine Just and Sara Dahlman (2023) that we must think beyond such binary stereotypes – not to achieve gender blindness, but to advance 'norm-critical' perspectives.

In the autoethnographic vignettes throughout this book, I have described many reasons why I was drawn to this study of public innovation in goods. In short, this research was motivated by a desire to change public policy.

The devices – the scientific instruments – were secondary. If anything, the 'geeky masculinity' that I catch myself performing would normally privilege governance over gadgets. Mine is a post-Captain Kirk, millennial generation, *Trekkie* masculinity. I am interpolated by the kind of cashless, classless, utopianism espoused by Gene Roddenberry in his *Star Trek* philosophy. I am further inspired by new *Trek*'s attempts to confront and deconstruct whiteness, colonialism, multiculturalism, and the gender binary. However, this is only a piece of my identity – all identities are 'fractured' (Haraway, 1990). At the time I overlooked Mann Borgese, I was also attempting to join in the epistemic culture of mainstream innovation studies, where a collective goal seems to be changing public policy in ways that only advance our *current* (dysfunctional) socioeconomic system. Inside the trappings of neoliberalized innovation studies, I felt like I was being assimilated by the Borg.

However, this book has demonstrated that was not the case. The Borg in *Star Trek* are drones whose technological appendages preclude individual thought and agency. They cannot drop their tools. We would be better to think of ourselves through Donna Harraway's notion of the cyborg. For her, 'a cyborg is a cybernetic organism, a hybrid of machine and organism, a creature of social reality as well as a creature of fiction' – and 'we are all cyborgs' (Haraway, 1990, p 191). Haraway argued that 'cyborg imagery can suggest a way out of the maze of dualisms in which we have explained our bodies and our tools to ourselves' (1990, p 223). Here, my instrumentalities and I are neither separate nor whole (you too have a role as reader). All told, there is a 'situatedness' to my study of public innovation in ocean science instruments. But I cannot simply *reflect* on my position and then neglect other possible knowledge positions. The cyborg metaphor is a call to refuse the 'allure' (Bell and Willmott, 2019) and 'safety' (Bochner, 2012) of feigning omniscience.

What we have, that the Borg do not, is the capacity for 'methodological humility' (Law and Singleton, 2005, p 350). We can inquire into our limits, and the limits of our tools, while accepting that we will never know it all. As I have just suggested, this is something more than methodological reflectivity. Law says that 'we cannot deconstruct all our subjectivities at the same time. And it may be that parts cannot be deconstructed at all' (1998, p 23). So, *after neoliberal tools and techniques*, we might begin to explore possibilities for 'disruptive reflexivity' (Bell and Willmott, 2019) or 'diffractive methodology' (Barad, 2007).

The first of these – disruptive reflexivity – is advocated by critical organizational scholars Emma Bell and Hugh Willmott (2019). It is a 'crafty' form of research practice that 'amplifies doubt by breaching convention and challenging the basis of knowledge claims' (Bell and Willmott, 2019, p 1370). It is about focusing on 'contingencies, paradoxes, and uncertainties in (social) scientific endeavour' (Bell and Willmott, 2019, p 1370). Bell and Willmott present this as a hallmark of 'intellectual craftship' (2019, p 1375), which they describe as 'an ethical and political, as well as skillful, embodied and imaginative activity' (2019, p 1380). They present it as a counterpoint to 'methodological fetishism' (2019, p 1375) and they incorporate Barad's (2007) ideas on entanglement.

Barad's diffractive methodology is the second major methodological direction that might advance a dark innovation agenda. Diffractive methodology is what Barad does when they read quantum physics and queer feminist philosophy 'through one another' (2007, p 30) – noticing the diffraction patterns. Barad argues 'that a diffractive methodology is respectful of the entanglement of ideas and other materials in ways that reflexive methodologies are not' (2007, p 29). This idea is itself entangled within the queer, feminist, decolonial theorizing of feminist new materialism (Harris and Ashcraft, 2023). It is one of many relational approaches that acknowledge how 'efforts to know a thing also enact that thing in particular ways' (Harris and Ashcraft, 2023, p 1). The method seeks creative rather than critical insights, through 'respectful engagements with different disciplinary practices' (Barad, 2007).

I built towards such approaches throughout this book, and yet there is still much work to be done. I have tried to *disrupt* innovation studies conventions but also to establish new *diffractive* patterns from the entanglements we call innovation studies, critical management studies, critical geography, science and technology studies, and ocean science. And like the high energy physicists, I have been primarily interested in establishing 'knowledge of the limits of knowing, of the mistakes we make in trying to know, of the things that interfere with our knowing, of what we are not interested in and do not really want to know' (Knorr Cetina, 1999, p 64). I have been obsessed with understanding how knowledge about innovation is disciplined through common tools and techniques.

This obsession means that I am concluding the book with more questions than answers. A reviewer told me that

> after all these twists and turns, I am not sure how much better I understand dark innovation. In fact, I feel like I have a better understanding of how difficult it is to beat back the darkness, and why innovation studies has opted against the path of thickets and thorns.

If you feel this same way, then the book has met its objective. It has shown that observing dark innovation will never be a leisurely day at the beach. I agree with Alf Rehn and Anders Örtenblad that what we need now is 'less clarity about innovation' and 'more challenges and debate' (2023, p 7). Innovation studies are entangled in an assemblage of *land technologies*. These help us cling to the safety of the shoreline. But dark innovation involves uncharted and unchartable waters – and that is where we must go if we are truly committed to the study of novelty.

Notes

Chapter 2

[1]   Although Shimshoni's thesis has multiple high-profile citations, the document itself appears lost to time. It is not available through Harvard's digital thesis repository and, despite help from some excellent library staff, I was unable to find a copy at any of the libraries that participate in WorldCat or in the Shimshoni collection at the Israeli National Library.

[2]   Printed with the permission of Eric von Hippel (2021). Personal communication to R. MacNeil, 28 September 2021.

[3]   While the triple helix model might have theoretical and empirical robustness (Shinn, 2002), I agree with Terry Shinn that it has barely registered in the literature. If it were more important for innovation studies, I would follow Pfotenhauer and Juhl (2017) and group the triple helix together with the systems of innovation approach as 'apolitical structures'. The triple helix is yet another systems model, with slightly different assumptions about academic, industry, and government roles.

Chapter 3

[1]   Dalhousie's cross-departmental oceanography 'institute' had recently achieved 'department' status, and this is noted with fanfare in the article.

[2]   The vessel was named for the famous arctic explorer Henry Hudson. Its initial designation CSS stood for 'Canadian Survey Ship'. This was an alternative to the designation given naval vessels of the era, Her Majesty's Canadian Ship or HMCS. After the establishment of the Canadian Coast Guard, the vessel would be restyled as the CCGS *Hudson* or Canadian Coast Guard Ship. See BIO (2002).

[3]   See Ørvik (1982), who suggests that the greatest threat to Canadian Arctic sovereignty was resource exploitation by American multinational corporations, not attack by Soviet submarines.

Chapter 4

[1]   'Canada's Hydrofoil, Revolutionary Idea for Navies', *LIFE*, 24 September 1954.

[2]   Pardon the pun, but the building was a new-to-Nova Scotia poured-concrete construction.

[3]   Again, pardon the pun.

[4]   Thanks to Archivist Rosemary Barbour for suggesting this perspective. In Chapter 8 it will become clear that I should have listened.

Chapter 5

[1]   Further extending the *incertae sedis* analogy to innovation taxonomies might help us think differently about the well-documented issues with high-level sectoral groupings of organizations like 'public sector' and 'service sector' (that is, *incertae familiae* or uncertain

family). Indeed, if 'the boundaries between manufacturing and services have blurred' (de Jong and Marsili, 2006, p 226), then the label 'services' creates waste basket taxa.

[2]   Here, my 'mare incognitum' analogy runs in parallel to the 'dark innovation' metaphor that drives this book. For a moment, physics and biology metaphors do not compete.

[3]   Note that I am not entering a debate about the 'length of the leash' between human biology and society/culture (cf. Nelson, 1995). Instead, I am following the broadly accepted conclusion that social life mediates all interactions with the material world – and vice-versa. The presence of any length leash disrupts the organism metaphor.

Chapter 6

[1]   Here, I echo Law and Singleton (2005).

References

Abernathy, W. J. and Utterback, J. M. (1978) 'Patterns of industrial innovation', *Technology Review*, 80(7), pp 40–7.

Adams, M., Sanderson, B., Porskamp, P., and Redden, A. M. (2019) 'Comparison of co-deployed drifting passive acoustic monitoring tools at a high flow tidal site: C-PODS and icListenHF hydrophones', *Journal of Ocean Technology*, 14, pp 59–83.

Afach, S., Buchler, B. C., Budker, D., Dailey, C., Derevianko, A., Dumont, V., et al (2021) 'Search for topological defect dark matter with a global network of optical magnetometers', *Nature Physics*, 17(12), pp 1396–401.

Agrawal, A., and Cockburn, I. (2003) 'The anchor tenant hypothesis: exploring the role of large, local, R&D-intensive firms in regional innovation systems', *International Journal of Industrial Organization*, 21(9), pp 1227–53.

Akrich, M. (1994) 'The de-scription of technical objects', in W. Bijker and J. Law (eds), *Shaping Technology/Building Society: Studies in Sociotechnical Change*. Cambridge, MA: MIT Press, pp 205–24.

Alcadipani, R. and Hassard, J. (2010) 'Actor-Network Theory, organizations and critique: towards a politics of organizing', *Organization*, 17(4), pp 419–35.

Allen, J. (2011a) 'Making space for topology', *Dialogues in Human Geography*, 1(3), pp 316–18.

Allen, J. (2011b) 'Topological twists: power's shifting geographies', *Dialogues in Human Geography*, 1(3), pp 283–98.

Almada Ventures Inc. (2013) *CinMaps.ca: Canadian Ocean Technology Map*. Available at: http://ocean.cinmaps.ca/home (Accessed: 22 February 2013).

Alonso, W., and Starr, P (1987) *The Politics of Numbers*. New York: Russell Sage Foundation.

Alvesson, M., and Sandberg, J. (2011) 'Generating research questions through problematization', *Academy of Management Review*, 36(2), pp 247–71.

Archibugi, D. (2001) 'Pavitt's Taxonomy Sixteen Years On: A Review Article', *Economics of Innovation and New Technology*, 10(5), pp 415–25.

Archibugi, D., and Filippetti, A. (2018) 'The retreat of public research and its adverse consequences on innovation', *Technological Forecasting and Social Change*, 127, pp 97–111.

Arundel, A., and Huber, D. (2013) 'From too little to too much innovation? Issues in measuring innovation in the public sector', *Structural Change and Economic Dynamics*, 27, pp 146–59.

Asdal, K. (2020) 'Is ANT equally good in dealing with local, national and global natures?', in A. Blok, I. Farias, and C. Roberts (eds), *The Routledge Companion to Actor-Network Theory*. Abingdon: Routledge, pp 337–44.

Asheim, B. T., and Coenen, L. (2005) 'Knowledge bases and regional innovation systems: comparing Nordic clusters', *Research Policy*, 34(8), pp 1173–90.

Asheim, B. T., and Isaksen, A. (2002) 'Regional innovation systems: the Integration of local "sticky" and global "ubiquitous" knowledge', *Journal of Technology Transfer*, 27(1), pp 77–86.

Asheim, B. T., Smith, H. L., and Oughton, C. (2011) 'Regional innovation systems: theory, empirics and policy', *Regional Studies*, 45(7), pp 875–91.

Atkinson-Grosjean, J. (2002) 'Canadian science at the public/private divide: the NCE experiment', *Journal of Canadian Studies*, 37(3), pp 71 –91.

Atkinson-Grosjean, J. (2006) *Public Science, Private Interests: Culture and Commerce in Canada's Networks of Centres of Excellence*. Toronto: University of Toronto Press.

Atlantic Coastal Zone Information Steering Committee (2006) *The Ocean Technology Sector in Atlantic Canada*. Moncton, NB: Atlantic Canada Opportunities Agency, Government of Canada.

Ausubel, J. H., Trew Crist, D. and Waggoner, P. E. (eds) (2010) *First Census of Marine Life 2010: Highlights of a Decade of Discovery*. Washington: Consortium for Ocean Leadership.

Backhaus, J. G. (2003) *Joseph Alois Schumpeter: Entrepreneurship, Style and Vision*. New York: Springer.

Bailey, M., Favaro, B., Otto, S. P., Charles, A., Devillers, R., Metaxas, A., et al (2016) 'Canada at a crossroad: the imperative for realigning ocean policy with ocean science', *Marine Policy*, 63, pp 53–60.

Baird, D. (2004) *Thing Knowledge*. Berkeley: University of California Press.

Ballestero, A. (2019) *A Future History of Water*. Durham, NC: Duke University Press.

Barabási, A. L. (2013) *Network Science*. Barabási Lab. Available at: http://net worksciencebook.com/ (Accessed: 31 October 2022).

Barad, K. (2007) *Meeting the Universe Halfway: Quantum Physics and the Entanglement of Matter and Meaning*. Durham, NC: Duke University Press.

Barnard, K. N. (1959) *Device for Holding Magnesium or Other Galvanic Anodes* (US Patent No. 3,012,959). Available at: https://image-ppubs. uspto.gov/dirsearch-public/print/downloadPdf/3012959 (Accessed: 31 October 2022).

Bascom, W. (1988) *The Crest of the Wave: Adventures in Oceanography*. New York: Harper & Row.

Becker, E. (1968) *The Structure of Evil: An Essay on the Unification of the Science of Man*. New York: G. Braziller.

Bell, Daniel (1973) *The Coming of Post-industrial Society: A Venture in Social Forecasting*. New York: Basic Books.

Bell, David (2013) 'Geek myths: technologies, masculinities, globalizations', in J. Hearn, M. Blagojević, and K. Harrison (eds), *Rethinking Transnational Men*. New York: Routledge, pp 76–90.

Bell, E., and Willmott, H. (2019) 'Ethics, politics and embodied imagination in crafting scientific knowledge', *Human Relations*, 73(10), pp 1366–87.

Belussi, F., Sammarra, A. and Sedita, S. R. (2010) 'Learning at the boundaries in an "Open Regional Innovation System": a focus on firms' innovation strategies in the Emilia Romagna life science industry', *Research Policy*, 39(6), pp 710–21.

Benson, K. R., and Rehbock, P. F. (1993) 'Oceanographic history: the Pacific and beyond', *Fifth International Congress on the History of Oceanography*, Scripps Institution of Oceanography, La Jolla, CA: University of Washington Press.

Berg, A.-J., and Lie, M. (1995) 'Feminism and Constructivism: Do Artifacts Have Gender?', *Science, Technology, & Human Values*, 20(3), pp 332–51.

Bijker, W. E., Hughes, T. P., and Pinch, T. J. (1987) *The Social Construction of Technological Systems: New Directions in the Sociology and History of Technology*. Cambridge, MA: MIT Press.

BIO (Bedford Institute of Oceanography) (1962–92) *Annual Reviews*. Dartmouth, NS: Government of Canada.

Black, E. (2001) *IBM and the Holocaust: The Strategic Alliance between Nazi Germany and America's Most Powerful Corporation*. New York: Crown Publishers.

Bochner, A. P. (2012) 'Suffering happiness: on autoethnography's ethical calling', *Qualitative Communication Research*, 1(2), pp 209–29.

Bochner, A. P. (2016) 'Putting meanings into motion: autoethnography's existential calling', in S. Holman Jones, T. E. Adams, and C. Ellis (eds), *Handbook of Autoethnography*. New York: Routledge, pp 50–6.

Bogers, M., Afuah, A., and Bastian, B. (2010) 'Users as innovators: a review, critique, and future research directions', *Journal of Management*, 36(4), pp 857–75.

Boileau, J. (2004) *Fastest in the World: The Saga of Canada's Revolutionary Hydrofoils*. Halifax, NS: Formac Publishing.

Boje, D. (2001) *Narrative Methods for Organizational & Communication Research*. London: Sage.

Boje, D., Rosile, G. A., Durant, R. A., and Luhman, J. T. (2004) 'Enron spectacles: a critical dramaturgical analysis', *Organization Studies*, 25(5), pp 751–74.

Boje, D. and Smith, R. (2010) 'Re-storying and visualizing the changing entrepreneurial identities of Bill Gates and Richard Branson', *Culture and Organization*, 16(4), pp 307–31.

Borgatti, S.P. (2002) *NetDraw Software for Network Visualization*. Lexington, KY: Analytic Technologies.

Borgatti, S. P. (2006) 'Identifying sets of key players in a social network', *Computational and Mathematical Organization Theory*, 12(1), pp 21–34.

Borgatti, S. P., and Molina, J.-L. (2005) 'Toward ethical guidelines for network research in organizations', *Social Networks*, 27(2), pp 107–17.

Borgatti, S. P., Everett, M. G., and Johnson, J. C. (2013) *Analyzing Social Networks*. London: Sage.

Borgese, E. M. (1973) *Pacem in Maribus*. New York: Dodd, Mead.

Bowker, G. C., and Star, S. L. (2000) *Sorting Things out: Classification and Its Consequences*. Cambridge, MA: MIT Press.

Boyce, D. G., Lewis, M. R., and Worm, B. (2010). 'Global phytoplankton decline over the past century', *Nature*, 466(7306), 591–6.

Brenner, N., Peck, J., and Theodore, N. (2010) 'Variegated neoliberalization: geographies, modalities, pathways', *Global Networks*, 10(2), pp 182–222.

Breschi, S. and Malerba, F. (1997) 'Sectoral innovation systems: technological regimes, Schumpeterian dynamics, and spatial boundaries', in C. Edquist (ed) *Systems of Innovation: Technologies, Institutions and Organizations*. Hoboken: Routledge, pp 130–56.

Breschi, S., Malerba, F., and Orsenigo, L. (2000) 'Technological regimes and Schumpeterian patterns of innovation', *Economic Journal*, 110(463), pp 388–410.

Brewer, D. D. (2000) 'Forgetting in the recall-based elicitation of personal and social networks', *Social Networks*, 22(1), pp 29–43.

Bromley, P., and Meyer, J. W. (2014) ' "They are all organizations": the cultural roots of blurring between the nonprofit, business, and government sectors', *Administration & Society*, 49(7), pp 939–66.

Bugge, M., Mortensen, P. S., and Bloch, C. (2011) *Measuring Public Innovation in Nordic Countries: Report on the Nordic Pilot Studies – Analyses of Methodology and Results*. Aarhus: Danish Centre for Studies in Research and Research Policy, Aarhus University.

Burns, T. and Stalker, G. M. (1961) *The Management of Innovation*. Oxford: Oxford University Press.

Bush, V. (1945) *Science, the Endless Frontier: A Report to the President*. Washington, DC: Government of the United States.

Butts, C. T. (2008) 'Social network analysis: a methodological introduction', *Asian Journal of Social Psychology*, 11(1), pp 13–41.

Calignano, G., Fitjar, R. D. and Kogler, D. F. (2018) 'The core in the periphery? The cluster organization as the central node in the Apulian aerospace district', *Regional Studies*, 52(11), pp 1490–1501.

Callaway, D. S., Newman, M. E., Strogatz, S. H., and Watts, D. J. (2000) 'Network robustness and fragility: percolation on random graphs', *Physical Review Letters*, 85(25). https://doi.org/10.1103/PhysRevLett.85.5468

Callon, M. (1986) 'The sociology of an actor-network: the case of the electric vehicle', in M. Callon, J. Law, and A. Rip (eds), *Mapping the Dynamics of Science and Technology: Sociology of Science in the Real World*. Basingstoke: Macmillan, pp 19–34.

Callon, M., and Law, J. (2005) 'On qualculation, agency, and otherness', *Environment and Planning D: Society and Space*, 23(5), pp 717–33.

Camus, A. (1955) *The Myth of Sisyphus*. London: Hamish Hamilton.

Canada with Names (2022). Ottawa: Natural Resources Canada. Available at: https://natural-resources.canada.ca/earth-sciences/geography/atlas-canada/explore-our-maps/reference-maps/16846 (Accessed: 31 October 2022).

'Canadian Institute of Oceanography' (1959) *Nature*, 184, p 769. https://doi.org/10.1038/184769c0

Cañibano, C., Encinar, M.-I., and Muñoz, F.-F. (2017) 'The economic rationality of NOvative behavior', in B. Godin and D. Vinck (eds), *Critical Studies of Innovation*. Cheltenham: Edward Elgar, pp 240–56.

Carroll, W. K. and Shaw, M. (2001) 'Consolidating a neoliberal policy bloc in Canada, 1976 to 1996', *Canadian Public Policy/Analyse de Politiques*, 27(2), pp 195–217.

Casper, S., Hollingsworth, J. R., and Whitley, R. (2005) 'Varieties of capitalism: comparative institutional approaches to economic organization and innovation', in S. Casper and F. van Waarden (eds), *Innovation and Institutions*. Cheltenham: Edward Elgar, pp 193–228.

Cassis, Y., and Minoglou, I. P. (2005) 'Entrepreneurship in theory and history: state of the art and new perspectives', in Y. Cassis and I. P. Minoglou (eds), *Entrepreneurship in Theory and History*. London: Palgrave Macmillan, pp 3–21.

Castellacci, F. (2008) 'Technological paradigms, regimes and trajectories: manufacturing and service industries in a new taxonomy of sectoral patterns of innovation', *Research Policy*, 37(6–7), pp 978–94.

Castellacci, F. (2009) 'The interactions between national systems and sectoral patterns of innovation', *Journal of Evolutionary Economics*, 19(3), pp 321–47.

Cesaratto, S. and Mangano, S. (1993) 'Technological profiles and economic performance in the Italian manufacturing sector', *Economics of Innovation and New Technology*, 2(3), pp 237–56.

Chaminade, C., and Plechero, M. (2015) 'Do regions make a difference? Regional innovation systems and global innovation networks in the ICT industry', *European Planning Studies*, 23(2), pp 215–37.

Chaturvedi, A. (2023) 'Non-Western perspectives on innovation', in A. Rehn and A. Örtenblad (eds), *Debating Innovation*. London: Palgrave Macmillan, pp 283–317.

Chidamber, S. R., and Kon, H. B. (1994) 'A research retrospective of innovation inception and success: the technology–push, demand–pull question', *International Journal of Technology Management*, 9(1), pp 94–112.

Clarke, A., Heffler, D., Belliveau, D., and Milligan, T. (2002) 'History of instrumentation at BIO', in Geddes, D. (ed), *Bedford Institute of Oceanography 2002 in Review*. Dartmouth, NS: Government of Canada, pp 37–42.

Clarke, A., Lazier, J., Petrie, B., Platt, T., Smith, P. and Elliott, J. (2002) 'Ocean sciences - looking back at 40 years', in Geddes, D. (ed), *Bedford Institute of Oceanography 2002 in Review*. Dartmouth, NS: Government of Canada, pp 24–7.

Coad, A., Nightingale, P., Stilgoe, J., and Vezzani, A. (2021) 'Editorial: the dark side of innovation', *Industry and Innovation*, 28(1), pp 102–12.

Coeckelbergh, M. (2015) *Money machines: Electronic Financial Technologies, Distancing, and Responsibility in Global Finance*. Abingdon: Routledge.

Coeckelbergh, M., and Reijers, W. (2016) 'Narrative technologies: a philosophical investigation of the narrative capacities of technologies by using Ricoeur's narrative theory', *Human Studies*, 39(3), pp 325–46.

Cohen, W. M. and Levinthal, D. A. (1994) 'Fortune favors the prepared firm', *Management Science*, 40(2), pp 227–51.

Cooke, B. (1999) 'Writing the left out of management theory: the historiography of the management of change', *Organization*, 6(1), pp 81–105.

Cooke, P. (2001) 'Regional innovation systems, clusters, and the knowledge economy', *Industrial and Corporate Change*, 10(4), pp 945–74.

Cooke, P. (2002) 'Biotechnology clusters as regional, sectoral innovation systems', *International Regional Science Review*, 25(1), pp 8–37.

Cooke, P. (2016) 'The virtues of variety in regional innovation systems and entrepreneurial ecosystems', *Journal of Open Innovation: Technology, Market, and Complexity*, 2, 13. https://doi.org/10.1186/s40852-016-0036-x

Coriat, B., and Weinstein, O. (2002) 'Organizations, firms and institutions in the generation of innovation', *Research Policy*, 31(2), pp 273–90.

Corrigan, L. T., and Mills, A. J. (2012) 'Men on board: Actor-Network Theory, feminism and gendering the past', *Management & Organizational History*, 7(3), pp 251–65.

Covin, J. G. and Miles, M. P. (1999) 'Corporate entrepreneurship and the pursuit of competitive advantage', *Entrepreneurship Theory and Practice*, 23(3), pp 47–63.

Cruz, S., Paulino, S., and Gallouj, F. (2015) 'Public service innovation: solid waste sector from the perspective of clean development mechanism landfill projects', *Journal of Inspiration Economy*, 2(2), pp 93–112.

Curnow, R. C., and Moring, G. G. (1968) ' "Project sappho": a study in industrial innovation', *Futures*, 1(2), pp 82–90.

Czarniawska, B. (2004) *Narratives in Social Science Research*. London: Sage.

D'Ignazio, C., and Klein, L. F. (2020) *Data Feminism*. Cambridge, MA: MIT Press.

Davis, G. F. (2022) *Taming Corporate Power in the 21st Century*. Cambridge: Cambridge University Press.

De Fuentes, C., and Dutrenit, G. (2012) 'Best channels of academia–industry interaction for long-term benefit', *Research Policy*, 41(9), pp 1666–82.

De Jong, J. P. J., and Marsili, O. (2006) 'The fruit flies of innovations: a taxonomy of innovative small firms', *Research Policy*, 35(2), pp 213–29.

De la Bellacasa, M. P. (2010) 'Matters of care in technoscience: assembling neglected things', *Social Studies of Science*, 41(1), pp 85–106.

De Laet, M., and Mol, A. (2000) 'The Zimbabwe bush pump: mechanics of a fluid technology', *Social Studies of Science*, 30(2), pp 225–63.

De Marchi, M., Napolitano, G., and Taccini, P. (1996) 'Testing a model of technological trajectories', *Research Policy*, 25(1), pp 13–23.

De Nooy, W., Mrvar, A., and Batagelj, V. (2011) *Exploratory Social Network Analysis with Pajek*. Cambridge: Cambridge University Press.

De Solla Price, D. J. (1984) 'The science/technology relationship, the craft of experimental science, and policy for the improvement of high technology innovation', *Research Policy*, 13(1), pp 3–20.

De Swart, J., Bertone, G., and van Dongen, J. (2017) 'How dark matter came to matter', *Nature Astronomy*, 1(3), pp 1–9.

De Vries, H., Bekkers, V., and Tummers, L. (2016) 'Innovation in the public sector: A systematic review and future research agenda', *Public Administration*, 94(1), pp 146–66.

Dekker, D., Krackhardt, D., and Snijders, T. A. B. (2007) 'Sensitivity of MRQAP tests to collinearity and autocorrelation conditions', *Psychometrika*, 72(4), pp 563–81.

Deleuze, G., and Guattari, F. (1987) *A Thousand Plateaus: Capitalism and Schizophrenia*. University of Minnesota Press.

Department of Oceanography (2011) *Making Waves: The Early Days of Oceanography at Dalhousie University*. Halifax, NS: Dalhousie University.

Derevianko, A., and Pospelov, M. (2014) 'Hunting for topological dark matter with atomic clocks', *Nature Physics*, 10(12), pp 933–6.

Derrida, J. (1978) *Writing and Difference*. Chicago: University of Chicago Press.

DiMaggio, P. J., and Powell, W. W. (1983) 'The iron cage revisited: institutional isomorphism and collective rationality in organizational fields', *American Sociological Review*, 48(2), pp 147–60.

Doern, G. B., Castle, D. and Phillips, P. W. B. (2016) *Canadian Science, Technology, and Innovation Policy*. Montreal: McGill-Queens University Press.

Doloreux, D., and Frigon, A. (2021) 'The innovation superclusters initiative in Canada: a new policy strategy?', *Science and Public Policy*, 49(1), pp 148–58.

Doloreux, D., and Parto, S. (2005) 'Regional innovation systems: current discourse and unresolved issues', *Technology in Society*, 27(2), pp 133–53.

Dosi, G. (1982) 'Technological paradigms and technological trajectories: A suggested interpretation of the determinants and directions of technical change', *Research Policy*, 11(3), pp 147–62.

DREA (ca. 1985) *Defence Research Establishment Atlantic*. Dartmouth, NS: Nova Scotia Public Archives Collection, V/F Vol 415 #11.

Durepos, G. (2015) 'ANTi-History: toward amodern histories', in P. Genoe McLaren, A. J. Mills, and T. Weatherbee (eds), *The Routledge Companion to Management and Organizational History*. New York: Routledge, pp 153–80.

Durepos, G., and Mills, A. J. (2011) 'Actor-Network Theory, ANTi-History and critical organizational historiography', *Organization*, 19(6), pp 703–21.

Durepos, G., and Mills, A. J. (2012) *ANTi-History: Theorizing the Past, History, and Historiography in Management and Organization Studies*. Charlotte, NC: Information Age Publishing.

Durepos, G., Mills, A. J. and Weatherbee, T. G. (2012) 'Theorizing the past: realism, relativism, relationalism and the reassembly of Weber', *Management & Organizational History*, 7(3), pp 267–81.

Durepos, G., Shaffner, E. C., and Taylor, S. (2021) 'Developing critical organizational history: context, practice and implications', *Organization*, 28(3), 449–67.

Dye, K., Mills, A. J., and Weatherbee, T. (2005) 'Maslow: man interrupted: reading management theory in context', *Management Decision*, 43(10), pp 1375–95.

Edquist, C. (1997) *Systems of Innovation: Technologies, Institutions, and Organizations*. Washington, DC: Pinter.

Edquist, C. (2001) 'The systems of innovation approach and innovation policy: an account of the state of the art', *DRUID*, Aalborg, Denmark, 12–15 June.

Edquist, C. (2004) 'Reflections on the systems of innovation approach', *Science and Public Policy*, 31(6), pp 485–9.

Edquist, C., and Johnson, B. (1997) 'Institutions and organisations in systems of innovation', in C. Edquist (ed), *Systems of Innovation: Technologies, Institutions and Organizations*. New York: Pinter, pp 41–63.

Ellis, C. (2004) *The Ethnographic I: A Methodological Novel about Autoethnography*. Lanham, MD: Rowman Altamira.

Fagerberg, J., Fosaas, M., Bell, M., and Martin, B. R. (2011) 'Christopher Freeman: social science entrepreneur', *Research Policy*, 40(7), pp 897–916.

Fagerberg, J., Fosaas, M., and Sapprasert, K. (2012a) 'Innovation: exploring the knowledge base', *Research Policy*, 41(7), pp 1132–53.

Fagerberg, J., and Verspagen, B. (2009) 'Innovation studies: the emerging structure of a new scientific field', *Research Policy*, 38(2), pp 218–33.

Fagerberg, J., Landström, H., and Martin, B. R. (2012b) 'Exploring the emerging knowledge base of "the knowledge society"', *Research Policy*, 41(7), pp 1121–31.

Fagerberg, J., Martin, B. R., and Andersen, E. S. (2013) *Innovation Studies: Evolution and Future Challenges*. Oxford: Oxford University Press.

Ferrary, M., and Granovetter, M. (2009) 'The role of venture capital firms in Silicon Valley's complex innovation network', *Economy and Society*, 38(2), pp 326–59.

Fløysand, A., and Jakobsen, S.-E. (2011) 'The complexity of innovation: a relational turn', *Progress in Human Geography*, 35(3), pp 328–44.

Fowler, G. A. (1997) *Wave-Powered Ocean Profiler* (US Patent No. 5,644,077). https://image-ppubs.uspto.gov/dirsearch-public/print/downloadPdf/5644 077 (Accessed: 31 October 2012).

Fowler, G.A., Hamilton, J. M., Beanlands, B. D., Belliveau, D. J., and Furlong, A. R. (1997). 'A wave powered profiler for long term monitoring', Oceans '97 MTS/IEEE Conference Proceedings, pp 225–8. https://doi.org/10.1109/OCEANS.1997.634366

Freeman, C. (1973) 'A study of success and failure in industrial innovation', in B. R. Williams (ed), *Science and Technology in Economic Growth: Proceedings of a Conference held by the International Economic Association at St Anton, Austria*. London: Palgrave Macmillan, pp 227–55.

Freeman, C. (1974) *Economics of Industrial Innovation*. London: Penguin.

Freeman, C. (1982) *The Economics of Industrial Innovation*. Cambridge, MA: MIT Press.

Freeman, C. (1987) *Technology, Policy, and Economic Performance: Lessons from Japan*. London: Pinter.

Freeman, C. (1991) 'Innovation, changes of techno-economic paradigm and biological analogies in economics', *Revue économique*, 42(2), pp 211–31.

Freeman, C. (1995) 'The "national system of innovation" in historical perspective', *Cambridge Journal of Economics*, 19(1), pp 5–24.

Freeman, C. (1996) 'The greening of technology and models of innovation', *Technological Forecasting and Social Change*, 53(1), pp 27–39.

Freeman, C. (2002) 'Continental, national and sub-national innovation systems – complementarity and economic growth', *Research Policy*, 31, pp 191–211.

Freeman, C., and Soete, L. (1997) *The Economics of Industrial Innovation*, 3rd ed. London: Pinter.

Fuenfschilling, L., and Truffer, B. (2014) 'The structuration of socio-technical regimes: conceptual foundations from institutional theory', *Research Policy*, 43(4), pp 772–91.

Funnell, L., and Dodds, K. (2016) *Geographies, Genders and Geopolitics of James Bond*. London: Palgrave Macmillan.

Gaddis, J. L. (2004) *The Landscape of History: How Historians Map the Past*. Oxford: Oxford University Press.

Galison, P. (1997) *Image and Logic: A Material Culture of Microphysics*. Chicago: University of Chicago Press.

Gallouj, F. (2002) *Innovation in the Service Economy: The New Wealth of Nations*. Cheltenham: Edward Elgar.

Gallouj, F., and Weinstein, O. (1997) 'Innovation in services', *Research Policy*, 26(4–5), pp 537–56.

Gallouj, F., and Zanfei, A. (2013) 'Innovation in public services: filling a gap in the literature', *Structural Change and Economic Dynamics*, 27, pp 89–97.

Garud, R., Gehman, J., and Giuliani, A. P. (2014) 'Contextualizing entrepreneurial innovation: a narrative perspective', *Research Policy*, 43(7), pp 1177–88.

Garud, R., Kumaraswamy, A., and Karnøe, P. (2010) 'Path dependence or path creation?', *Journal of Management Studies*, 47(4), pp 760–74.

Gault, F. (2012) 'User innovation and the market', *Science and Public Policy*, 39(1), pp 118–28.

Gault, F. (2018) 'Defining and measuring innovation in all sectors of the economy', *Research Policy*, 47(3), pp 617–22.

Gault, F. (2020) *Measuring Innovation Everywhere*. Cheltenham: Edward Elgar.

Gay, B., and Dousset, B. (2005) 'Innovation and network structural dynamics: study of the alliance network of a major sector of the biotechnology industry', *Research Policy*, 34(10), pp 1457–75.

Genoe McLaren, P., Mills, A. J., and Durepos, G. (2009) 'Disseminating Drucker', *Journal of Management History*, 15(4), pp 388–403.

Gephart, R. P. (1988) *Ethnostatistics: Qualitative Foundations for Quantitative Research*. Newbury Park, CA: Sage.

Gephart, R. P. (1997) 'Hazardous measures: an interpretive textual analysis of quantitative sensemaking during crises', *Journal of Organizational Behavior*, 18(S1), pp 583–622.

Gephart, R. P. (2006) 'Ethnostatistics and organizational research methodologies: an introduction', *Organizational Research Methods*, 9(4), pp 417–31.

Gereffi, G., Brun, L. C., Lee, J., and Turnipseed, M. (2013) *Nova Scotia's Ocean Technologies: A Global Value Chain Analysis of Inshore & Extreme Climate Vessels, Remotely Operated & Autonomous Underwater Vehicles, and Underwater Sensors & Instrumentation*. Durham, NC: Duke University Press.

Gertler, M. S. (2010) 'Rules of the game: the place of institutions in regional economic change', *Regional Studies*, 44(1), pp 1–15.

Ghent, J. M. (1981) 'Cooperation in science and technology', in A. Balawyder (ed), *Canadian-Soviet Relations 1939–1980*. Oakville, ON: Mosaic Press, pp 173–92.

Godin, B. (2002) 'The rise of innovation surveys: measuring a fuzzy concept', *Canadian Science and Innovation Indicators Consortium, Project on the History and Sociology of S&T Statistics*, Working Paper No. 16.

Godin, B. (2005) *Measurement and Statistics on Science and Technology: 1920 to the Present*. Abingdon: Routledge.

Godin, B. (2006) 'The linear model of innovation: the historical construction of an analytical framework', *Science, Technology, & Human Values*, 31(6), pp 639–67.

Godin, B. (2008) 'In the shadow of Schumpeter: W. Rupert Maclaurin and the study of technological innovation', *Minerva*, 46(3), pp 343–60.

Godin, B. (2011) 'The linear model of innovation: Maurice Holland and the research cycle', *Social Science Information*, 50(3–4), pp 569–81.

Godin, B. (2012) '"Innovation studies": the invention of a specialty', *Minerva*, 50(4), pp 397–421.

Godin, B. (2014) '"Innovation studies": staking the claim for a new disciplinary "tribe"', *Minerva*, 52(4), pp 489–95.

Godin, B. (2017) *Models of Innovation: The History of an Idea.* Cambridge, MA: MIT Press.

Godin, B. (2019) *The Invention of Technological Innovation: Languages, Discourses and Ideology in Historical Perspective.* Cheltenham: Edward Elgar.

Godin, B. (2020) *The Idea of Technological Innovation: A Brief Alternative History.* Cheltenham: Edward Elgar.

Godin, B., and Lane, J. P. (2013) 'Pushes and pulls: hi(S)tory of the demand pull model of innovation', *Science, Technology, & Human Values*, 38(5), pp 621–54.

Godin, B., and Vinck, D. (2017a) *Critical Studies of Innovation: Alternative Approaches to the Pro-innovation Bias.* Cheltenham: Edward Elgar.

Godin, B., and Vinck, D. (2017b) 'Introduction: innovation – from the forbidden to a cliché', in B. Godin and D. Vinck (eds), *Critical Studies of Innovation.* Cheltenham: Edward Elgar, pp 1–14.

Goldstein, P. Z. (1997) 'How many things are there? A reply to Oliver and Beattie, Beattie and Oliver, Oliver and Beattie, and Oliver and Beattie', *Conservation Biology*, 11(2), pp 571–4.

Gorm Hansen, B. (2011) 'Beyond the boundary: science, industry, and managing symbiosis', *Bulletin of Science, Technology & Society*, 31(6), pp 493–505.

Government of Nova Scotia (2012) *Defined by the Sea: Nova Scotia's Oceans Technology Sector Present and Future.* Halifax, NS: Government of Nova Scotia.

Greater Halifax Partnership (2012) *Ocean Technology.* Halifax, NS. http://www.greaterhalifax.com/en/home/halifax-nova-scotia/business-sectors/ocean_industries/ocean_technology.aspx (Accessed: 31 October 2012).

Grodal, S. (2017) 'Field expansion and contraction: how communities shape social and symbolic boundaries', *Administrative Science Quarterly*, pp 0001839217744555.

Grønning, T. (2008) 'Institutions and innovation systems: the meanings and roles of the institution concept within systems of innovation approaches' *DRUID*, Copenhagen, June.

Grosser, T. J., Lopez-Kidwell, V., and Labianca, G. (2010) 'A social network analysis of positive and negative gossip in organizational life', *Group & Organization Management*, 35(2), pp 177–212.

Guildline (1973). Salinometer Prototype [scientific instrument]. Ottawa: Ingenium – Canada's Museums of Science and Innovation. http://collection.ingeniumcanada.org/en/item/2017.0006.001/

Gupta, V. (2009) 'On black elephants'. 27 April 2009. http://vinay.howtol ivewiki.com/blog/flu/on–black–elephants–1450 2021.

Halvorsen, T., Hauknes, J., Miles, I., and Røste, R. (2005) *On the Differences between Public and Private Innovation*. Oslo: PUBLIN. http://unpan1.un.org/intradoc/groups/public/documents/apcity/unpan046809.pdf (Accessed: 1 November 2013).

Hamblin, J. D. (2002) 'Environmental diplomacy in the Cold War: the disposal of radioactive waste at sea during the 1960s', *International History Review*, 24(2), pp 348–75.

Hamblin, J. D. (2005) *Oceanographers and the Cold War: Disciples of Marine Science*. Seattle, WA: University of Washington Press.

Hamblin, J. D. (2008) *Poison in the Well: Radioactive Waste in the Oceans at the Dawn of the Nuclear Age*. New Brunswick, NJ: Rutgers University Press.

Hannan, M. T. and Freeman, J. (1977) 'The population ecology of organizations', *American Journal of Sociology*, 82(5), pp 929–64.

Hanneman, R. A., and Riddle, M. (2005) *Introduction to Social Network Methods*. Riverside, CA: University of California, Riverside.

Hansen, M. B. (2011) 'Antecedents of organizational innovation: the diffusion of new public management into Danish local government', *Public Administration*, 89(2), pp 285–306.

Haraway, D. (1988) 'Situated knowledges: the science question in feminism and the privilege of partial perspective', *Feminist Studies*, 14(3), pp 575–99.

Haraway, D. (1990) 'A manifesto for cyborgs: science, technology, and socialist feminism in the 1980s', in L. J. Nicholson (ed), *Feminism/Postmodernism*. New York: Routledge, pp 190–233.

Harris, K. L., and Ashcraft, K. L. (2023). 'Deferring difference no more: an (im)modest, relational plea from/through Karen Barad', *Organization Studies*. https://doi.org/10.1177/01708406231169424

Hartt, C. M. (2013) 'Actants without actors: polydimensional discussion of a regional conference', *Tamara Journal for Critical Organization Inquiry*, 11(3), pp 15–25.

Hartt, C. M. (2019) *Connecting Values to Action: Non-corporeal Actants and Choice*. Bingley: Emerald.

Hartt, C. M., Mills, A. J., Helms Mills, J., and Corrigan, L. T. (2014) 'Sense-making and actor networks: the non-corporeal actant and the making of an Air Canada history', *Management & Organizational History*, 9(3), pp 288–304.

Helmreich, S. (2009) *Alien Ocean: Anthropological Voyages in Microbial Seas*. Berkeley: University of California Press.

Helms Mills, J., Weatherbee, T. G., and Colwell, S. R. (2006) 'Ethnostatistics and sensemaking: making sense of university and business school accreditation and rankings', *Organizational Research Methods*, 9(4), pp 491–515.

Hoag, H. (2011) 'Canadian research shift makes waves', *Nature*, 472, p 269.

Hoag, H. (2012) 'Canadian budget hits basic science', *Nature News*, 30 March.

Hoag, H. (2013) 'Lady of the lakes', *Nature*, 502(7473), pp 612–13.

Hoberg, G., and Phillips, G. (2016) 'Text-based network industries and endogenous product differentiation', *Journal of Political Economy*, 124(5), pp 1423–65.

Hodgson, G. M. (2002) 'Darwinism in economics: from analogy to ontology', *Journal of Evolutionary Economics*, 12(3), pp 259–81.

Holbrook, J. A. and Wolfe, D. A. (2000) *Innovation, Institutions and Territory: Regional Innovation Systems in Canada*. Montréal: McGill-Queen's University Press.

Hollenstein, H. (2003) 'Innovation modes in the Swiss service sector: a cluster analysis based on firm-level data', *Research Policy*, 32(5), pp 845–63.

Hood, C. (1991) 'A public management for all seasons?', *Public Administration*, 69(1), pp 3–19.

Howard, C. (2016) 'Neoliberal numbers: calculation and hybridization in Australian and Canadian official statistics', in B. Michelle and K. L. Randy (eds), *Governing Practices*. Toronto: University of Toronto Press, pp 131–54.

Hubert, L. J. (1987) *Assignment Methods in Combinatorial Data Analysis*. New York: Dekker.

Hughes, R. I. G. (1997) 'Models and representation', *Philosophy of Science*, 64(S4), pp S325–S36.

Hughes, T. P. (1969) 'Technological momentum in history: hydrogenation in Germany 1898–1933', *Past & Present*, 44, pp 106–32.

Hughes, T. P. (1976) *Science and the Instrument-Maker: Michelson, Sperry, and the Speed of Light*. Washington, DC: Smithsonian Institution Press.

Huisman, M. (2009) 'Imputation of missing network data: some simple procedures', *Journal of Social Structure*, 10(1), pp 1–29.

Hyysalo, S., Pollock, N., and Williams, R. A. (2019) 'Method matters in the social study of technology: investigating the biographies of artifacts and practices', *Science & Technology Studies*, 32(3), pp 2–25.

Industry Canada (2013) *Canadian Company Capabilities*. Ottawa, ON. http://www.ic.gc.ca/eic/site/ccc-rec.nsf/eng/home (Accessed: 1 November 2013).

Inglott, P. S. (2004) 'Elisabeth Mann Borgese: metaphysician by birth', *Ocean Yearbook*, 18(1), pp 22–74.

'Institute of Oceanography, Dalhousie: Prof Ronald Hayes' (1959) *Nature*, 183, pp 1161. https://doi.org/10.1038/1831161a0

International Ocean Institute (n.d.) *The Founder of IOI*. Malta. https://www.ioinst.org/elisabeth-mann-borgese/ (Accessed: 1 December 2022).

Irvine, J., and Martin, B. R. (1984a) 'CERN: Past performance and future prospects: II. The scientific performance of the CERN accelerators', *Research Policy*, 13(5), pp 247–84.

Irvine, J., and Martin, B. R. (1984b) *Foresight in Science: Picking the Winners*. London: Pinter.

Irwin, A., Vedel, J. B., and Vikkelsø, S. (2021) 'Isomorphic difference: familiarity and distinctiveness in national research and innovation policies', *Research Policy*, 50(4). https://doi-org.ezproxy.acadiau.ca:9443/10.1016/j.respol.2021.104220

Jacques, R. (2006) 'History, historiography and organization studies: the challenge and the potential', *Management & Organizational History*, 1(1), pp 31–49.

Jensen, T., and Sandström, J. (2020) 'Organizing rocks: Actor-Network Theory and space', *Organization*, 27(5), pp 701–16.

Joerges, B., and Shinn, T. (2001) *Instrumentation between Science, State and Industry*. New York: Springer.

Just, S. N. and Dahlman, S. (2023) 'What does it take? Feminist readings of innovation studies', in A. Rehn and A. Örtenblad (eds) *Debating Innovation: Perspectives and Paradoxes of an Idealized Concept*. New York: Springer, pp 263–82.

Kelley, E. S., Mills, A. J., and Cooke, B. (2006) 'Management as a Cold War phenomenon?', *Human Relations*, 59(5), pp 603–10.

Kennedy, D. P., and McCarty, C. (2016) *EgoWeb 2.0* (computer software). http://github.com/qualintitative/egoweb (Accessed: 18 September 2023).

Kennedy, H., and Hill, R. L. (2018) 'The feeling of numbers: emotions in everyday engagements with data and their visualisation', *Sociology*, 52(4), pp 830–48.

Khaire, M. (2014) 'Fashioning an industry: socio-cognitive processes in the construction of worth of a new industry', *Organization Studies*, 35(1), pp 41–74.

Kilduff, M. (1992) 'The friendship network as a decision-making resource: dispositional moderators of social influences on organizational choice', *Journal of Personality and Social Psychology*, 62(1), pp 168–80.

Kilduff, M., and Oh, H. (2006) 'Deconstructing diffusion: an ethnostatistical examination of medical innovation network data reanalyses', *Organizational Research Methods*, 9(4), pp 432–55.

Kirsch, D., Moeen, M., and Wadhwani, R. (2014) 'Historicism and industry emergence: industry knowledge from pre-emergence to stylized fact', in R. D. Wadhwani and M. Bucheli (eds), *Organizations in Time: History, Theory, Methods*. Oxford University Press, pp 217–40.

Kline, S. J. (1985) 'Innovation is not a linear process', *Research Management*, 28(4), pp 36–45.

Kline, S. J., and Rosenberg, N. (1986) 'An overview of innovation', in R. Landau and N. Rosenberg (eds), *The Positive Sum Strategy: Harnessing Technology for Economic Growth*. Washington, DC: National Academy Press, pp 275–305.

Knorr Cetina, K. (1999) *Epistemic Cultures: How the Sciences Make Knowledge*. Cambridge, MA: Harvard University Press.

Koch, P., and Hauknes, J. (2005) *On Innovation in the Public Sector*. Oslo: Nordic Institute for Studies in Innovation, Research and Education.

Krackhardt, D. (1988) 'Predicting with networks: nonparametric multiple regression analysis of dyadic data', *Social Networks*, 10(4), pp 359–81.

Kuhn, T. S. (1962) *The Structure of Scientific Revolutions*. Chicago: University of Chicago Press.

Landström, H., and Lohrke, F. (2010) *Historical Foundations of Entrepreneurial Research*. Cheltenham: Edward Elgar.

Langrish, J. (1974) 'The changing relationship between science and technology', *Nature*, 250(5468), pp 614–16.

Langrish, J. (2017) 'Physics or biology as models for the study of innovation', in B. Godin and D. Vinck (eds), *Critical Studies of Innovation*. Cheltenham: Edward Elgar, pp 296–318.

Langrish, J., Gibbons, M., Evans, W. G. and Jevons, F. R. (1972) *Wealth from Knowledge: Studies of Innovation in Industry*. London: Palgrave Macmillan.

Lata, I. B., and Minca, C. (2016) 'The surface and the abyss/rethinking topology', *Environment and Planning D*, 34(3), pp 438–55.

Latham, A. (2011) 'Topologies and the multiplicities of space-time', *Dialogues in Human Geography*, 1(3), pp 312–15.

Latour, B. (1987) *Science in Action: How to Follow Scientists and Engineers through Society*. Cambridge, MA: Harvard University Press.

Latour, B. (1992) 'Where are the missing masses? The sociology of a few mundane artifacts', in W. Bijker and J. Law (eds), *Shaping Technology/ Building Society: Studies in Sociotechnical Change*. Cambridge, MA: MIT Press, pp 205–24.

Latour, B. (1993) *The Pasteurization of France*, translated by A. Sheridan and J. Law. Cambridge, MA: Harvard University Press.

Latour, B. (1999) 'On recalling ANT', in J. Law and J. Hassard (eds), *Actor Network Theory and after*. Oxford: Blackwell, pp 15–25.

Latour, B. (2005) *Reassembling the Social: An Introduction to Actor-Network Theory*. Oxford: Oxford University Press.

Latour, B., and Woolgar, S. (1986) *Laboratory Life: The Construction of Scientific Facts*. Princeton: Princeton University Press.

Law, J. (1986) *Power, Action, and Belief: A New Sociology Of Knowledge?* London: Routledge.

Law, J. (1994) *Organizing Modernity*. Cambridge, MA: Blackwell.

Law, J. (1998) 'Machinic pleasures and interpellations', in B. Brenna, J. Law and I. Moser (eds), *Machines, Agency and Desire*. Oslo: Centre for Technology and Culture, pp 23–48.

Law, J. (1999) 'After ANT: complexity, naming and topology', *The Sociological Review*, 47(1), pp 1–14.

Law, J. (2004) *After Method: Mess in Social Science Research*. New York: Routledge.

Law, J. (2016) 'STS as method', in U. Felt, R. Fouché, C. Miller, and L. Smith-Doerr (eds), *The Handbook of Science and Technology Studies*. Cambridge, MA: MIT Press, pp 31–58.

Law, J., and Hassard, J. (1999) *Actor Network Theory and after*. Malden, MA: Blackwell/Sociological Review.

Law, J., and Mol, A. (2001) 'Situating technoscience: an inquiry into spatialities', *Environment and Planning D*, 19(5), pp 609–21.

Law, J., and Singleton, V. (2005) 'Object lessons', *Organization*, 12(3), pp 331–55.

Lawrence, T. B., and Suddaby, R. (2006) 'Institutions and institutional work', in S. R. Clegg, T. B. Lawrence, and C. Hardy (eds), *The Sage Handbook of Organization Studies*. London: Sage, pp 215–54.

Lawson, B.T. (2023). *The Life of a Number: Measurement, Meaning and the Media*. Bristol: Bristol University Press.

Leiponen, A. and Drejer, I. (2007) 'What exactly are technological regimes?: Intra-industry heterogeneity in the organization of innovation activities', *Research Policy*, 36(8), pp 1221–38.

Leitner, K.-H. (2017) ' "No" and "slow" innovation strategies as a response to increased innovation speed', in B. Godin and D. Vinck (eds), *Critical Studies of Innovation*. Cheltenham: Edward Elgar, pp 201–18.

Levallois, C. (2011) 'Why were biological analogies in economics "a bad thing"? Edith Penrose's battles against Social Darwinism and McCarthyism', *Science in Context*, 24(4), pp 465–85.

Lewis, M. R., and Smith, J. C. (1983). 'A small volume, short-incubation-time method for measurement of photosynthesis as a function of incident irradiance', *Marine Ecology – Progress Series*, 13(1), 99–102.

Liebowitz, S. J., and Margolis, S. E. (1995) 'Path dependence, lock-in, and history', *Journal of Law, Economics, & Organization*, 11(1), pp 205–26.

Linnaeus, C. (1758) *Systema Naturæ*. Stockholm: Laurentius Salvius.

Lippert, I. (2018) 'On not muddling lunches and flights: narrating a number, qualculation, and ontologising troubles', *Science & Technology Studies*, 31(4), pp 52–74.

Lippert, I., and Verran, H. (2018) 'After numbers? Innovations in science and technology studies' analytics of numbers and numbering', *Science & Technology Studies*, 31(4), pp 2–12.

List, F. (1841) *The National System of Political Economy*. London: Longmans, Green, and Company.

Liu, H., and Pechenkina, E. (2019) 'Innovation-by-numbers: an autoethnography of innovation as violence', *Culture and Organization*, 25(3), pp 178–88.

Lofting, H. (1920) *The Story of Doctor Dolittle*. New York: Frederick A. Stokes.

Longard, J. R. (1993) *Knots, Volts and Decibels: An Informal History of the Naval Research Establishment, 1940–1967*. Dartmouth, NS: Defence Research Establishment Atlantic.

Lopez-Kidwell, V. (2013) 'The heart of social networks: the ripple effect of emotional abilities in relational well-being'. PhD thesis, University of Kentucky.

Lorenz, C. (2012) 'If you're so smart, why are you under surveillance? Universities, neoliberalism, and new public management', *Critical Inquiry*, 38(3), pp 599–629.

Louçã, F., and Cabral, R. (2021) 'Chris Freeman's concept of evolution: a critique of the misuse of biological analogies in macroeconomics', *Research Policy*, 50(9). https://doi.org/10.1016/j.respol.2021.104322

Lundvall, B.-Å. (1988) 'Innovation as an interactive process: from user-producer interaction to the national system of innovation', in G. Dosi, C. W. Freeman, R. R. Nelson, and L. Soete (eds), *Technical Change and Economic Theory*. London: Pinter, pp 349–69.

Lundvall, B.-Å. (1992) *National Systems of Innovation: Towards a Theory of Innovation and Interactive Learning*. New York: Pinter.

Lundvall, B.-Å. (2013a) 'An agenda for future research', in J. Fagerberg, B. R. Martin, and E. S. Andersen (eds), *Innovation Studies: Evolution and Future Challenges*. Oxford: Oxford University Press, pp 202–9.

Lundvall, B.-Å. (2013b) 'Innovation studies: a personal interpretation of "the state of the art"', in J. Fagerberg, B. R. Martin, and E. S. Andersen (eds), *Innovation Studies: Evolution and Future Challenges*. Oxford: Oxford University Press, pp 21–71.

Lundvall, B.-Å. (2016) 'From manufacturing nostalgia to a strategy for economic transformation', *Economia e Politica Industriale*, 43(3), pp 265–71.

Lyotard, J.-F. (1984) *The Postmodern Condition: A Report on Knowledge*. Minneapolis: University of Minnesota Press.

MacKenzie, D. A., and Wajcman, J. (1999) *The Social Shaping of Technology*. Milton Keynes: Open University Press.

Maclaurin, W. R. (1949) *Invention and Innovation in the Radio Industry*. Cambridge, MA: MIT Press.

Maclaurin, W. R. (1950) 'The process of technological innovation: the launching of a new scientific industry', *American Economic Review*, 40(1), pp 90–112.

MacNeil, R. T. (2014) 'Hybrid, larval and symbiotic companies: some taxonomic problems in Nova Scotia's ocean technology industries'. *Proceedings of the Atlantic Schools of Business*, Mount Saint Vincent University, Halifax, NS, pp 69–87.

MacNeil, R. T. (2019) '"We are the tools": an ANTi-History unboxing of the "boys' toys challenge" for innovation studies', *11th International Conference in Critical Management Studies*, Milton Keynes, 27 June.

MacNeil, R. T., and Mills, A. J. (2015) 'Organizing a precarious black box: an actor–network account of the Atlantic Schools of Business, 1980–2006', *Canadian Journal of Administrative Sciences/Revue Canadienne des Sciences de l'Administration*, 32(3), pp 203–13.

MacNeil, R. T., Ochoa Briggs, S., Christie, A. E., and Sheehan, C. (2021) 'Beyond the ecosystem metanarrative: narrative multiplicity and entrepreneurial experiences at the University of Waterloo', in P. Eriksson, U. Hytti, K. Komulainen, T. Montonen, and P. Siivonen (eds), *New Movements in Academic Entrepreneurship*. Cheltenham: Edward Elgar, pp 83–103.

Malerba, F. (2005) 'Sectoral systems of innovation: a framework for linking innovation to the knowledge base, structure and dynamics of sectors', *Economics of Innovation and New Technology*, 14(1–2), pp 63–82.

Mandel, E. (1978) *Late Capitalism*. London: Verso.

Mann Borgese, E. (1998) *The Oceanic Circle: Governing the Seas as a Global Resource*. Tokyo: United Nations University Press.

Mannheim, K. (1936) *Ideology and Utopia*. London: Routledge.

Marcacci, F. (2019) 'Illusions, ghosts and movies in the history of scientific instruments' in V. Fano (ed), *Gino Tarozzi Philosopher of Physics*. Milan: FrancoAmgeli, pp 148–60.

Marshall, A. (1890) *Principles of Economics*. London: Macmillan.

Martin, B. (2010) *Science Policy Research: Having an Impact on Policy?* London: Office of Health Economics.

Martin, B. (2013) 'Innovation studies: an emerging agenda', in J. Fagerberg, B. R. Martin, and E. S. Andersen (eds), *Innovation Studies: Evolution and Future Challenges*. Oxford: Oxford University Press, pp 168–93.

Martin, B. (2016) 'Twenty challenges for innovation studies', *Science and Public Policy*, 43(3), pp 432–50.

Martin, B., Nightingale, P., and Yegros-Yegros, A. (2012) 'Science and technology studies: exploring the knowledge base', *Research Policy*, 41(7), pp 1182–204.

Martin, B., Salter, A., and Hicks, D. (1996) *The Relationship between Publicly Funded Basic Research and Economic Performance*. Brighton: University of Sussex, Science Policy Research Unit.

Martin, B. R., and Irvine, J. (1984a) 'CERN: past performance and future prospects: I. CERN's position in world high-energy physics', *Research Policy*, 13(4), pp 183–210.

Martin, B. R., and Irvine, J. (1984b) 'CERN: past performance and future prospects: III. CERN and the future of world high-energy physics', *Research Policy*, 13(6), pp 311–42.

Martin, J. L. (1999) 'A general permutation-based QAP analysis approach for dyadic data from multiple groups', *Connections*, 22(2), pp 50–60.

Mazzucato, M. (2013) *The Entrepreneurial State: Debunking Public vs. Private Myths in Innovation*. London: Anthem Press.

Mazzucato, M. (2016) 'From market fixing to market-creating: a new framework for innovation policy', *Industry and Innovation*, 23(2), pp 140–56.

McLaren, P. G., and Durepos, G. (2019) 'A call to practice context in management and organization studies', *Journal of Management Inquiry*, 30(1), pp 74–84.

McLeod, P. (2011) 'NDP effort 'waste of money'; MacKay slams $1.4-million Ships Start Here lobbying campaign', *The Chronicle Herald*, 20 October 2011, p A6.

Meeus, M., and Oerlemans, L. (2005) 'National innovation systems', in S. Casper and F. van Waarden (eds), *Innovation and Institutions: A Multidisciplinary Review of the Study of Innovation Systems*. Cheltenham: Edward Elgar, pp 152–89.

Meyer, T. (2022) *Elisabeth Mann Borgese and the Law of the Sea*. Leiden: Brill.

Mills, A. J. (2010) 'Juncture', in A. J. Mills, G. Durepos, and E. Wiebe (eds), *Encyclopedia of Case Study Research*. London: Sage, pp 509–11.

Mills, E. L. (1993) 'Pacific waters and the POG: the origin of physical oceanography on the west coast of Canada', in K. R. Benson and P. F. Rehbock (eds), *Oceanographic History: The Pacific and beyond*. La Jolla, CA: University of Washington Press, pp 303–15.

Mills, E. L. (1994) 'Bringing oceanography into the Canadian university classroom', *Scientia Canadensis: Canadian Journal of the History of Science, Technology and Medicine*, 18(1), pp 3–21.

Mills, E. L. (2011a) *The Fluid Envelope of Our Planet: How the Study of Ocean Currents Became a Science*. Toronto: University of Toronto Press.

Mintzberg, H. (1996) 'Managing government, governing management', *Harvard Business Review*, 74(3), pp 75–85.

Mintzberg, H. (2015) 'Time for the plural sector', *Stanford Social Innovation Review* (Summer 2015), pp 28–33.

Mokyr, J. (2013) 'Capitalism reinvents itself', *Current History*, 112(757), pp 291–7.

Mol, A. (2002) *The Body Multiple: Ontology in Medical Practice*. Durham, NC: Duke University Press.

Mol, A., and Law, J. (1994) 'Regions, networks and fluids: anaemia and social topology', *Social Studies of Science*, 24(4), pp 641–71.

Mol, A., and Law, J. (2004) 'Embodied action, enacted bodies: the example of hypoglycaemia', *Body & Society*, 10(2–3), pp 43–62.

Monaghan, S. (1974). 'Watery wonder of the world', *Dalhousie Gazette*, 107(5), 10 October, p 1.

Moore, M. H. (2005) 'Break-through innovations and continuous improvement: two different models of innovative processes in the public sector', *Public Money & Management*, 25(1), pp 43–50.

Morgan, A. (2014) 'The rise of the geek: exploring masculine identity in The Big Bang Theory', *Masculinities*, 2, pp 31–57.

Morgan, G. (1980) 'Paradigms, metaphors, and puzzle solving in organization theory', *Administrative Science Quarterly*, 25(4), 605–22.

Morgan, G. (1986) *Images of Organization*. Beverly Hills, CA: Sage.

Morgan, G. (1997) *Images of Organization*, 2nd ed. Thousand Oaks, CA: Sage.

Mowery, D. C. (2009) 'National security and national innovation systems', *Journal of Technology Transfer*, 34(5), pp 455–73.

Mowery, D. C. and Nelson, R. R. (1996) 'The US corporation and technical progress', in Kaysen, C. (ed) *The American Corporation Today*. Oxford: Oxford University Press, pp 187–241.

Mowery, D. C., and Rosenberg, N. (1979) 'The influence of market demand upon innovation: a critical review of some recent empirical studies', *Research Policy*, 8(2), pp 102–53.

Myers, S., and Marquis, D. G. (1969) *Successful Industrial Innovations: A Study of Factors Underlying Innovation in Selected Firms*. Washington, DC: National Science Foundation.

Myrick, K., Helms Mills, J., and Mills, A. J. (2013) 'History-making and the Academy of Management: an ANTi-History perspective', *Management & Organizational History*, 8(4), pp 345–70.

Ndiaye, P. (2007) *Nylon and Bombs: DuPont and the March of Modern America*. Baltimore: Johns Hopkins University Press.

Nelson, R. R. (1993) *National Innovation Systems: A Comparative Analysis*. Oxford: Oxford University Press.

Nelson, R. R. (1995) 'Recent evolutionary theorizing about economic change', *Journal of Economic Literature*, 33(1), pp 48–90.

Nelson, R. R., and Winter, S. G. (1982) *An Evolutionary Theory of Economic Change*. Cambridge, MA: Belknap Press of Harvard University Press.

Nemet, G. F. (2009) 'Demand-pull, technology-push, and government-led incentives for non-incremental technical change', *Research Policy*, 38(5), pp 700–9.

Nichols, B. (2002) 'A short history of the Bedford Institute of Oceanography', in Geddes, D. (ed), *Bedford Institute of Oceanography 2002 in Review*. Dartmouth, NS: Government of Canada, pp 14–19.

Niosi, J., and Zhegu, M. (2005) 'Aircraft systems of innovation', in J. Niosi (ed), *Canada's Regional Innovation Systems*. Montreal: McGill-Queen's University Press, pp 61–86.

Niosi, J., and Zhegu, M. (2010) 'Anchor tenants and regional innovation systems: the aircraft industry', *International Journal of Technology Management*, 50(3/4), pp 263–82.

Nooteboom, B. (2000) 'Institutions and forms of co-ordination in innovation systems', *Organization Studies*, 21(5), pp 915–39.

North, D. C. (1990) *Institutions, Institutional Change and Economic Performance.* Cambridge: Cambridge University Press.

NSRF (1946–95) *Annual Reports of the Nova Scotia Research Foundation.* Nova Scotia Public Archives: Call Number J 104 .K3 .R29 R432.

O'Doherty, D. P. (2013). 'Off-road and spaced-out in the city: organization and the interruption of topology', *Space and Culture*, 16(2), 211–28.

O'Doherty, D. P. (2023). 'The animal spirits of innovation: on companion species, creativity, and Olly the airport cat', in A. Rehn and A. Örtenblad (eds), *Debating Innovation.* London: Palgrave Macmillan, pp 357–87.

O'Dor, R. K., Andrade, Y., Webber, D. M., Sauer, W. H. H., Roberts, M. J., Smale, M. J., et al (1998) 'Applications and performance of Radio-Acoustic Positioning and Telemetry (RAPT) systems', *Hydrobiologia*, 371–72, pp 1–8.

Ocean Technology Council of Nova Scotia (2013) *Member Directory.* http://www.otcns.ca/index.php?option=com_content&view=article&id=48&Itemid=58 (Accessed: 24 October 2013).

OECD (Organisation for Economic Co-operation and Development) (2005) *The Measurement of Scientific and Technological Activities, Oslo Manual Guidelines for Collecting and Interpreting Innovation Data.* Paris: OECD.

Oliver, N. and Blakeborough, M. (1998) 'Innovation Networks: The View from the Inside', in J. Michie and J. G. Smith (eds), *Globalization, Growth, and Governance: Creating an Innovative Economy.* Oxford: Oxford University Press, pp 146–60.

Oppenheim, R. (2020) 'How does the South Korean city of Kyŏngju help ANT think place and scale?', in A. Blok, I. Farias and C. Roberts (eds), *The Routledge Companion to Actor-Network Theory.* London: Routledge, pp 318–27.

Ørvik, N. (1982) *Canada's Northern Security: The Eastern Dimension.* Kingston, ON: Centre for International Relations, Queen's University.

Paasi, A. (2011) 'Geography, space and the re-emergence of topological thinking', *Dialogues in Human Geography*, 1(3), pp 299–303.

Palmer, D. (2006) 'Taking stock of the criteria we use to evaluate one another's work: ASQ 50 years out', *Administrative Science Quarterly*, 51(4), pp 535–59.

Parker, B. R. (2005) *Death Rays, Jet Packs, Stunts & Supercars: The Fantastic Physics of Film's Most Celebrated Secret Agent.* Baltimore: Johns Hopkins University Press.

Pavitt, K. (1984) 'Sectoral patterns of technical change: towards a taxonomy and a theory', *Research Policy*, 13(6), pp 343–73.

Peck, J. (2010a) *Constructions of Neoliberal Reason*. Oxford: Oxford University Press.

Peck, J. (2010b) 'Zombie neoliberalism and the ambidextrous state', *Theoretical Criminology*, 14(1), pp 104–10.

Peck, J. and Theodore, N. (2019) 'Still neoliberalism?', *South Atlantic Quarterly*, 118(2), pp 245–65.

Peck, J. and Tickell, A. (2002) 'Neoliberalizing space', *Antipode*, 34(3), pp 380–404.

Peck, J., Brenner, N., and Theodore, N. (2018) 'Actually existing neoliberalism', in D. Cahill, M. Cooper, M. Konings, and D. Primrose (eds), *The Sage Handbook of Neoliberalism*. London: Sage, pp 3–15.

Peneder, M. (2003) 'Industry classifications: aim, scope and techniques', *Journal of Industry, Competition and Trade*, 3(1–2), pp 109–29.

Peneder, M. (2010) 'Technological regimes and the variety of innovation behaviour: creating integrated taxonomies of firms and sectors', *Research Policy*, 39(3), pp 323–34.

Penrose, E. T. (1952) 'Biological analogies in the theory of the firm', *American Economic Review*, 42(5), pp 804–19.

Penrose, E. T. (1959) *The Theory of the Growth of the Firm*. Oxford: Oxford University Press.

Perani, G. (2019) 'Business innovation statistics and the evolution of the Oslo Manual', *NOvation: Critical Studies of Innovation*, I, pp 135–70.

Perani, G. (2021) 'Business innovation measurement: history and evolution', in B. Godin, G. Gaglio, and D. Vinck (eds), *Handbook on Alternative Theories of Innovation*. Cheltenham: Edward Elgar, pp 292–308.

Perry, J. L., and Rainey, H. G. (1988) 'The public-private distinction in organization theory: a critique and research strategy', *Academy of Management Review*, 13(2), pp 182–201.

Peter, S., and Lawrence, T. C. (2017) 'ANTi-History and the entrepreneurial work of privateers', *Qualitative Research in Organizations and Management: An International Journal*, 12(2), pp 94–110.

Pfotenhauer, S., and Juhl, J. (2017) 'Innovation and the political state: beyond the myths of technologies and markets', in B. Godin and D. Vinck (eds), *Critical Studies of Innovation*. Cheltenham: Edward Elgar, pp 68–93.

Phelan, S. (2007) 'Messy grand narrative or analytical blind spot? When speaking of neoliberalism', *Comparative European Politics*, 5(3), pp 328–38.

Phelan, S. (2014) *Neoliberalism, Media and the Political*. New York: Springer.

Pigott, P. (2011) *From Far and Wide: A Complete History of Canada's Arctic Sovereignty*. Toronto: Dundurn Press.

Polanyi, M. (1962) 'The Republic of Science: its political and economy theory', *Minerva*, 1(1), pp 54–73.

Popper, K. R. (2005) *The Logic of Scientific Discovery*. London: Routledge.

Porac, J. F., Thomas, H., and Baden-Fuller, C. (2011) 'Competitive groups as cognitive communities: the case of Scottish knitwear manufacturers revisited', *Journal of Management Studies*, 48(3), pp 646–64.

Porac, J. F., Thomas, H., Wilson, F., Paton, D., and Kanfer, A. (1995) 'Rivalry and the industry model of Scottish knitwear producers', *Administrative Science Quarterly*, pp 203–27.

Porter, M. (1990) *Competitive Advantage of Nations*. New York: The Free Press.

Porter, M. (2003) 'The economic performance of regions', *Regional Studies*, 37(6–7), pp 545–546.

Powell, W. W., Packalen, K., and Whittington, K. (2012) 'Organizational and institutional genesis', in J. F. Padgett and W. W. Powell (eds), *The Emergence of Organizations and Markets*. Princeton: Princeton University Press, pp 434–65.

Prasad, A. (2002) 'The contest over meaning: hermeneutics as an interpretive methodology for understanding texts', *Organizational Research Methods*, 5(1), pp 12–33.

Prasad, A. (2019) *Autoethnography and Organization Research*. New York: Springer.

Redden, A. M. (2016) 'Engaging locally and globally to address tidal energy research challenges & youth education', *Marine Renewables Canada*, 3–4 November.

Rehn, A. (2023). 'Image, imperatives, and ideology in the innovation industry', in A. Rehn and A. Örtenblad (eds), *Debating Innovation*. London: Palgrave Macmillan, pp 77–99.

Rhodes, C. (2021) *Woke Capitalism: How Corporate Morality Is Sabotaging Democracy*. Bristol: Policy Press.

Ricoeur, P. (1984) *Time and Narrative*, translated by K. McLaughlin and D. Pellauer, Chicago: University of Chicago Press.

Riggs, W., and von Hippel, E. (1994) 'Incentives to innovate and the sources of innovation: the case of scientific instruments', *Research Policy*, 23(4), pp 459–69.

Ritvo, H. (1997) *The Platypus and the Mermaid, and Other Figments of the Classifying Imagination*. Cambridge, MA: Harvard University Press.

Rose, N. (1991) 'Governing by numbers: figuring out democracy', *Accounting, Organizations and Society*, 16(7), pp 673–92.

Rosenberg, N. (1969) 'The direction of technological change: inducement mechanisms and focusing devices', *Economic Development and Cultural Change*, 18(1), pp 1–24.

Rosenberg, N. (1982) *Inside the Black Box*. Cambridge: Cambridge University Press.

Rosenberg, N. (1992) 'Scientific instrumentation and university research', *Research Policy*, 21(4), pp 381–90.

Rosenthal, R. (1979) 'An introduction to the file drawer problem', *Psychological Bulletin*, 86, pp 638–41.

Røste, R. (2005) *Studies of Innovation in the Public Sector: A Literature Review*. Oslo: PUBLIN.

Rothwell, R., Freeman, C., Horlsey, A., Jervis, V. T. P., Robertson, A. B., and Townsend, J. (1974) 'SAPPHO updated – project SAPPHO phase II', *Research Policy*, 3(3), pp 258–91.

Roundy, P. T. (2016) 'Start-up community narratives: the discursive construction of entrepreneurial ecosystems', *Journal of Entrepreneurship*, 25(2), pp 232–48.

Roundy, P. T. (2018) 'Rust belt or revitalization: competing narratives in entrepreneurial ecosystems', *Management Research Review*, 42(1), pp 102–21.

Roundy, P. T., and Bayer, M. A. (2018) 'Entrepreneurial ecosystem narratives and the micro-foundations of regional entrepreneurship', *International Journal of Entrepreneurship and Innovation*, 20(1). https://doi.org/10.1177/1465750318808426

Sá, C. M. (2022) 'The curious story of the Global Innovation Clusters renewal', *University Affairs*. https://www.affairesuniversitaires.ca/opinion/policy-and-practice/the-curious-story-of-the-global-innovation-clusters-renewal/ (Accessed: 18 September 2023).

Sá, C. M., and Litwin, J. (2011) 'University-industry research collaborations in Canada: the role of federal policy instruments', *Science and Public Policy*, 38(6), pp 425–35.

Saifer, A., and Dacin, M. T. (2021) 'Data and organization studies: aesthetics, emotions, discourse and our everyday encounters with data', *Organization Studies*, 43(4), pp 623–36.

Salazar, M., and Holbrook, J. A. (2007) 'Canadian science, technology and innovation policy: the product of regional networking?', *Regional Studies*, 41(8), pp 1129–41.

Salter, A. J., and Martin, B. R. (2001) 'The economic benefits of publicly funded basic research: a critical review', *Research Policy*, 30(3), pp 509–32.

Sandberg, J., and Alvesson, M. (2011) 'Ways of constructing research questions: gap-spotting or problematization?', *Organization*, 18(1), pp 23–44.

Sanderson, B. G., Stokesbury, M. J., and Redden, A. M. (2021) 'Using trajectories through a tidal energy development site in the Bay of Fundy to study interaction of renewable energy with local fish', *Journal of Ocean Technology*, 16(1), pp 55–74.

Sardar, Z. and Sweeney, J. A. (2016) 'The three tomorrows of postnormal times', *Futures*, 75, pp 1–13.

Sartre, J.-P., and Richmond, S. (1956) *Being and Nothingness: An Essay in Phenomenological Ontology*. London: Routledge.

Schmookler, J. (1966) *Invention and Economic Growth*. Cambridge, MA: Harvard University Press.

Schubert, T. (2009) 'Empirical observations on New Public Management to increase efficiency in public research – boon or bane?', *Research Policy*, 38(8), pp 1225–34.

Schumpeter, J. (1934) *The Theory of Economic Development*. New Brunswick, NJ: Transaction Books.

Sepp, V. (2012) 'Inoperative provinces, immobile regions and the geography of heterogeneous associations: The case of absent territorial border change in Estonia', *Geografiska Annaler: Series B, Human Geography*, 94(1), pp 47–63.

Shimshoni, D. L. (1966) 'Aspects of scientific entrepreneurship'. PhD thesis, Harvard University.

Shimshoni, D. L. (1970) 'The mobile scientist in the American instrument industry', *Minerva*, 8(1), pp 59–89.

Shinn, T. (2002) 'The triple helix and new production of knowledge: prepackaged thinking on science and technology', *Social Studies of Science*, 32(4), pp 599–614.

Shinn, T. (2005) 'New sources of radical innovation: research-technologies, transversality and distributed learning in a post-industrial order', *Social Science Information*, 44(4), pp 731–64.

Sinclair, M., Stobo, W., Lavoie, R., O'Boyle, B., and Marshall, L. (2002) 'Research in support of fisheries', in Geddes, D. (ed), *Bedford Institute of Oceanography 2002 in Review*. Dartmouth, NS: Government of Canada, pp 30–4.

Smith, D. (2004) 'A decade of doing things differently: universities and public-sector reform in Manitoba', *Canadian Public Administration*, 47(3), pp 280–303.

Smith, W. L., Boje, D. M., and Gardner, C. (2004) 'Using the ethnostatistics methodology to reconcile rhetoric and reality: an examination of the management release of Enron's year end 2000 results', *Qualitative Research in Accounting & Management*, 1(2), pp 1–16.

Snelgrove, P. V. (2010) *Discoveries of the Census of Marine Life: Making Ocean Life Count*. Cambridge: Cambridge University Press.

Soete, L. (2019) 'Science, technology and innovation studies at a crossroad: SPRU as case study', *Research Policy*, 48(4), pp 849–57.

Soltis, S. M. (2012) 'Who you are and who you know: the influence of person-environment fit and social network centrality on individual performance'. PhD thesis, University of Kentucky.

Spital, F. C. (1979) 'An analysis of the role of users in the total R&D portfolios of scientific instrument firms', *Research Policy*, 8(3), pp 284–96.

Stork, D., and Richards, W. D. (1992) 'Nonrespondents in communication network studies', *Group & Organization Management*, 17(2), pp 193–209.

Suddaby, R., Foster, W. M., and Trank, C. Q. (2010) 'Rhetorical history as a source of competitive advantage', *Advances in Strategic Management*, 27, pp 147–74.

Takeda, Y., Kajikawa, Y., Sakata, I., and Matsushima, K. (2008) 'An analysis of geographical agglomeration and modularized industrial networks in a regional cluster: a case study at Yamagata prefecture in Japan', *Technovation*, 28(8), pp 531–9.

Taleb, N. N. (2010) *The Black Swan: The Impact of the Highly Improbable*. New York: Random House.

Tang, X., Wang, L., and Kishore, R. (2014) 'Why do IS scholars cite other scholars? An empirical analysis of the direct and moderating effects of cooperation and competition among IS scholars on individual citation behavior', *International Conference on Information Systems*, Auckland, New Zealand.

Taub, L. (2011) 'Introduction: reengaging with Instruments', *Isis*, 102(4), pp 689–96.

Taub, L. (2019) 'What is a scientific instrument, now?', *Journal of the History of Collections*, 31(3), pp 453–67.

Trenbirth, D. (1960) 'Canada becomes base for big sea study', *The Chronicle Herald*, 6 August 1960, p 6.

Trist, E. L. and Bamforth, K. W. (1951) 'Some social and psychological consequences of the longwall method of coal-getting', *Human Relations*, 4(1), pp 3–38.

Tureta, C., Américo, B. L., and Clegg, S. (2021) 'Controversies as method for ANTi-History: an inquiry into public administration practices', *Organization*, 28(6), pp 1018–35.

Turner, C. (2013) *The War on Science: Muzzled Scientists and Wilful Blindness in Stephen Harper's Canada*. Vancouver: Greystone Books.

Utterback, J. M. (1969) 'The process of technical innovation in instrument firms'. PhD thesis, MIT.

Utterback, J. M. (1971a) 'The process of innovation: a study of the origination and development of ideas for new scientific instruments', *IEEE Transactions on Engineering Management*, 18(4), pp 124–31.

Utterback, J. M. (1971b) 'The process of technological innovation within the firm', *Academy of Management Journal*, 14(1), pp 75–88.

Utterback, J. M. (1974) 'Innovation in industry and the diffusion of technology', *Science*, 183(4125), pp 620–6.

Vaara, E., Sonenshein, S., and Boje, D. (2016) 'Narratives as sources of stability and change in organizations: approaches and directions for future research', *Academy of Management Annals*, 10(1), pp 495–560.

Van Steenburgh, W. E. (1962) Address at the Special Convocation at Dalhousie University on the Occasion of the Official Opening of the Bedford Institute of Oceanography. Halifax, NS: Dalhousie University. http://www.bio-oa.ca/bioexpo/docs/VanSteenburg-address1962.pdf

Vermeulen, N. (2013) 'From Darwin to the census of marine life: marine biology as big science', *PLoS ONE*, 8(1), https://doi.org/10.1371/journal.pone.0054284

Verran, H. (2001) *Science and an African Logic*. Chicago: University of Chicago Press.

Verspagen, B. and Werker, C. (2004) 'Keith Pavitt and the invisible college of the economics of technology and innovation', *Research Policy*, 33(9), pp 1419–31.

Vinsel, L., and Russell, A. L. (2020) *The Innovation Delusion: How Our Obsession with the New Has Disrupted the Work That Matters Most.* New York: Currency.

von Hippel, E. (1975) 'The dominant role of users in the scientific instrument innovation process'. PhD thesis, MIT.

von Hippel, E. (1976) 'The dominant role of users in the scientific instrument innovation process', *Research Policy*, 5(3), pp 212–39.

von Hippel, E. (1986) 'Lead users: a source of novel product concepts', *Management Science*, 32(7), pp 791–805.

von Hippel, E. (1988) *The Sources of Innovation*. Oxford: Oxford University Press.

Vromen, J. J. (2006) 'Routines, genes and program-based behavior', *Journal of Evolutionary Economics*, 16(5), pp 543–60.

Wadhwani, R. D., and Bucheli, M. (2014) 'The future of the past in management and organization studies', in M. Bucheli. and R. D. Wadhwani (eds), *Organizations in Time: History, Theory, Methods.* Oxford: Oxford University Press, pp 3–32.

Wadhwani, R. D., and Jones, G. (2014) 'Schumpeter's plea: historical reasoning in entrepreneurship theory and research', in M. Bucheli. and R. D. Wadhwani (eds), *Organizations in Time: History, Theory, Methods.* Oxford: Oxford University Press, pp 192–216.

Wadhwani, R. D., and Lubinski, C. (2017) 'Reinventing entrepreneurial history', *Business History Review*, 91(4), pp 767–99.

Waite, P. B. (1994) *The Lives of Dalhousie University*. Montreal: McGill-Queen's University Press.

Walker, R. M. (2014) 'Internal and external antecedents of process innovation: a review and extension', *Public Management Review*, 16(1), pp 21–44.

Walker, R. M., Jeanes, E., and Rowlands, R. (2002) 'Measuring innovation – applying the literature-based innovation output indicator to public services', *Public Administration*, 80(1), pp 201–14.

Warner, D. J. (1990) 'What is a scientific instrument, when did it become one, and why?', *The British journal for the history of science*, 23(1), pp 83–93.

Watkins, L. (1980) 'Halifax-Dartmouth area: one of the three biggest marine science centres in Western Hemisphere', *Canadian Geographic*, 100, pp 12–23.

Weick, K. E. (1996) 'Drop your tools: an allegory for organizational studies', *Administrative Science Quarterly*, 41, pp 301–13.

Whelan, E. (2001) 'Politics by other means: feminism and mainstream science studies', *Canadian Journal of Sociology/Cahiers canadiens de sociologie*, 26(4), pp 535–81.

White, H. V. (1987) *The Content of the Form: Narrative Discourse and Historical Representation*. Baltimore: Johns Hopkins University Press.

Wikipedia (2017) 'Q (James Bond)'. *"Wikipedia, The Free Encyclopedia"*. https://en.wikipedia.org/wiki/Q_(James_Bond)

Williams, B. R. (1973) *Science and Technology in Economic Growth*. London: Palgrave Macmillan.

Windrum, P., and Koch, P. M. (2008) *Innovation in Public Sector Services: Entrepreneurship, Creativity and Management*. Cheltenham: Edward Elgar.

Winner, L. (1980) 'Do artifacts have politics?', *Daedalus*, 109(1), pp 121–36.

Winner, L. (2018) 'The cult of innovation: its myths and rituals', in E. Subrahmanian, T. Odumosu, and J. Y. Tsao (eds), *Engineering a Better Future: Interplay between Engineering, Social Sciences, and Innovation*. New York: Springer, pp 61–73.

Witt, U. (1996) 'A "Darwinian revolution" in economics?', *Journal of Institutional and Theoretical Economics*, 152(4), pp 707–15.

Woese, C. R., Kandler, O., and Wheelis, M. L. (1990) 'Towards a natural system of organisms: proposal for the domains Archaea, Bacteria, and Eucarya', *Proceedings of the National Academy of Sciences*, 87(12), pp 4576–9.

Wolfe, D. A. (2014) *Innovating in Urban Economics: Economic Transformation in Canadian City-Regions*. Toronto: University of Toronto Press.

Woodward, J. (1965) *Industrial Organization: Theory and Practice*. Oxford: Oxford University Press.

Woolgar, S., Coopmans, C. and Neyland, D. (2009) 'Does STS mean business? ', *Organization*, 16(1), pp 5–30.

Yin, R. K. (2009) *Case study research: design and methods*. Los Angeles: Sage.

Ziman, J. (2003a) 'Evolutionary models for technological change', in J. Ziman (ed), *Technological Innovation as an Evolutionary Process*. Cambridge: Cambridge University Press, pp 3–12.

Ziman, J. (2003b) 'Selectionism and complexity', in J. Ziman (ed), *Technological Innovation as an Evolutionary Process*. Cambridge: Cambridge University Press, pp 41–51.

Ziman, J. (2003c) *Technological Innovation as an Evolutionary Process*. Cambridge: Cambridge University Press.

Index

References to endnotes show both the page number and the note number (149n1).